VICTORIA AND ALBERT MUSEUM

IRONWORK

PART III.—A COMPLETE SURVEY OF THE
ARTISTIC WORKING OF IRON IN GREAT
BRITAIN FROM THE EARLIEST TIMES

J. STARKIE GARDNER

ILLUSTRATED

LONDON : PRINTED UNDER THE AUTHORITY
OF HIS MAJESTY'S STATIONERY OFFICE. 1922

© Crown copyright 1978

First published 1922

Photolitho impression with
Supplementary Bibliography compiled by Marian Campbell
1978

isbn 0 905209 02 8

Front cover: *Fragment from a gate. Showing the emblems of the United
Kingdom. Wrought iron. English, 18th century. 501–1901*

Back cover: *Mask from Hampton Court Palace Fountain Garden Gates.
Wrought iron. Made by Jean Tijou, c. 1690. Lent by H.M. Ministry of
Works.*

PREFATORY NOTE.

THIS publication forms the third and concluding volume of the Handbook on Ironwork by Mr. J. Starkie Gardner. It has been considered desirable, in the interest of readers, to include the whole range of English ironwork within a single volume. The early part of the book therefore goes over ground already more fully covered in "Ironwork, Part I.," and some of the illustrations included in that volume appear again here. The material provided by Mr. Gardner has been prepared for, and seen through, the Press by Mr. W. W. Watts, and the other Officers of the Department of Metalwork.

Thanks are due to Messrs. B. T. Batsford, Ltd., for their kind permission to reproduce several illustrations from two of their publications.

CECIL H. SMITH.

Victoria and Albert Museum.
December 1921.

CONTENTS.

LIST OF ILLUSTRATIONS.

CHAP. I.—THE PRODUCTION OF IRON IN BRITAIN.

IN olden times the great centres of iron production were the Forest Regions. Gloucester and other towns round about the Forest of Dean vied with Nottingham, Sheffield, and the towns of Derbyshire situated in the wooded region of which Sherwood Forest formed the centre; while these were rivalled by the iron-forges of the Weald of Sussex and of Kent.

Of these the Forest of Dean claims first consideration. Not only was it the chief Iron Country of the Romans in Britain, but it held the premier position throughout mediæval history. With a circuit of about 30 miles between the Severn and the Wye, its ores rich, abundant, easily accessible and near the surface, it combined a supply of fuel which for centuries appeared well-nigh inexhaustible.

In Roman times, and even earlier, the exports must have been large, though unfortunately Pliny speaks only in general terms of the iron-smelting of Britain. The trade was probably in the possession of the Basques, or other of the many races acquainted with iron, who may have found refuge here in prehistoric times. Our gleaming cliffs must, times out of number, have appeared a land of safety to peoples driven west by conquering hordes, which Asia seemed ever ready to pour forth, resulting in the parcelling out of our Island by many tribes under different rulers.

From the direction of the Forest of Dean came the so-called " Currency " bars, a convenient form for merchantable iron, adapted for transport and ready for use. These are believed to date back to the second century B.C., and are found as they were abandoned or lost, sometimes in considerable quantities, the finds radiating east from the Forest of Dean, or its neighbourhood. It remained the chief iron-field under the Romans from Gallienus to the close of the third century A.D. As elsewhere in the vast Roman Empire, the mines were worked systematically by driving galleries and sinking pits, the cinders from their smelting in which Roman coins are found covering many miles. Vast caverns also remain from which the ore has been extracted. The many iron objects of Roman date found in the Forest were doubtless local productions. As the Romans only knew the foot-blast, the cinders left by them contain 40 per cent. of iron ; the vast heaps, Cinder Hill for instance, solid masses of scoriæ, having thus proved a source of wealth for centuries. Yielding more kindly metal than the virgin ores, they have been preferred and worked uninterruptedly to the special benefit of the nearest city, Gloucester. This soon became an important manufacturing centre, and the only one mentioned in Domesday as liable to a tribute of iron bars for the Navy, to which it furnished anchors, cables, bolts, etc., as it did in previous centuries, possibly to the ships of the Veneti. Gloucester also developed a trade in nails, horse-shoes, and so on, for Henry II. requisitioned 25,000 nails to be sent from Gloucester to his palace at Winchester. Henry III. commanded the sheriff to repair the castle wall overlooking the town forges. These were in the hands of nobles and men of wealth : in 1140 the Abbot of Flaxley was empowered to erect forges. Giraldus Cambrensis is among writers who note the celebrity of Gloucester for its iron manufactures. It remained the Sheffield of its time, and held monopolies such as the supply of iron, coal, and other metals to Ireland. Reverting to the Forest of Dean, Leland in the time of Henry VIII. remarks that " the soil is fruitful of iron mines and divers forges be there to

make it " ; Camden speaks of the old cinders as producing the best qualities of iron ; and Musket tells us that
the numerous blast-furnaces had been supplied from this
source for 300 years; Atkyns, in his history of Gloucestershire, sets out the advantage accruing to landowners from
selling the cinders at good prices to the furnaces. Pepys
visited the Forest and notes the value of these vast heaps
of cinders, which he says are necessary for the making of
iron and without which men cannot work. The monopolies
of the Free Miners were not abolished till 1676. Soon after,
in 1687, when the best qualities of iron were made from
the cinders and beaten into bars of various dimensions
for market, regulations were issued including the cost of
carriage. The failing wood supply eventually caused the
industry to languish, and by 1720 there remained but
ten furnaces in the Forest. Revival, however, rapidly
followed the use of coal for smelting, since in 1795, 225,000
tons were being produced annually from only eight furnaces ; still chiefly from cinders, for less than half the
quantity was obtained from quarried hæmatite.

Roman forges and ironworks have also been found at
Weston-under-Panyard, Herefordshire, the Roman Arico-
nium ; and the remarkable discovery of three iron pigs
weighing from 256 to 484 lb. in the Roman villa at Chad-
worth, shows that they were able to smelt it. Cinders, foot-
blasts, and Roman coins have been found as far off as
Worcester. The Romans also worked iron in the Mendip
and Brendon Hills, and on Exmoor ; in Derbyshire,
Weardale, Cleveland, and near York ; and in Northumberland, Cumberland, and Durham. It is curious to note that
at Lanchester the blast is said to have been produced by
converging tunnels in the hill facing west. Still more remarkable is the Roman recognition, identified by the presence of coins in the slag, of the value of the rich ores in

Northamptonshire, their works for smelting being found at King's Cliff, Bulwick, and Rockingham, just below the surface. When the forests were used up, the ores were forgotten, and not rediscovered till 1800.

The ores of the Weald of Sussex were unknown to the Britons, and only became known to the Romans when well established here. Coins have been found in slag at Maresfield, Sedlescombe, Westfield, Buxted, and Crowborough, but other than a few of Nero's time, they date from about 200 years later. The ores were forgotten, and the Andredeswalde relapsed into the impenetrable forest implied in its name, without roads or navigable waters, until 1226, when the Weald once more comes into notice, through the grant by Henry II. to Lewes of a penny for every cart and a halfpenny for every horse loaded with iron passing through the town.

A steelyard weight with the arms of Richard, King of the Romans, has been found in Lewes Castle. Nevertheless, Henry II. sent to Gloucester for nails for his palace at Winchester, as before stated, and a bishop of Chichester also ordered his iron to be sent from there. In 1300 the London Corporation, on complaints of short measure, sealed three patterns of iron sent from the Weald for wheelwrights; and in 1320, 29,000 nails and 3,000 horse-shoes were requisitioned from the Weald for the Scottish Wars; while in 1342 tithes were levied on iron. The Weald prospered, and in 1436 there were about 140 hammers; until, the wood beginning to fail, we find a prohibition to erect new furnaces. The desire of Henry VIII. for greatly increased quantities of guns and ammunition led to the casting of cannon and shot in the Weald, when he sent his Italian and other bronzefounders to assist Huggett and John Owen in this undertaking. An immense export trade set in soon after Henry's death, when many highly placed persons embarked in the iron trade. A census of Sussex Ironworks in 1574 included Queen Elizabeth and George Boleyn, the Earl of Northumberland, Lord Abergavenny, Sir Thomas Gresham, Sir John Pelham, Sir Henry Sidney, and many others. The Lord High Admiral when impeached owned several furnaces. Count Gondomar often begged leave to export big guns, which were the staple—Thomas Johnson, who died about 1600, being the best founder. Firebacks cast with the arms of most of the owners are in existence. Elizabeth, however, taxed iron, coal, or ore, passing over the roads of the Weald, at a shilling per load, and the price continued to rise till it was cheaper to import from abroad. Far-seeing men like Harrison and Fuller began to lament that coals could not be sent into Sussex or to Southampton, and the foundries had to be worked with charcoal, then becoming unprocurable. Camden describes the Weald as full of iron mines in sundry places and furnaces on every side with " a huge deal of wood " burnt yearly, while Norden states there were still 140 furnaces and hammers there. Founders had migrated, as Leland tells us, and introduced the industry into Glamorganshire, where Ralfes, Hodgsons and Morleys, Sussex names, are still found. Iron was smelted at Aberdare and Merthyr Tydfil with wood while it lasted. Under the Commonwealth the blast-furnaces of the Weald had dwindled to 27 and the forges to 42, and in 1740 these had fallen to 14, in 1790 to 2 ; and the last, that of Ashburnham, was extinguished in 1828, when the local smiths were paying £24 per ton for the iron.

The destruction of wood brought iron-founding to an end in many places besides the Weald, as in Shropshire and the Mendips, and in Wales. Fuller exclaims, " Oh if this coale could be so charked as to make iron melt out of the stone as it maketh it in smiths' forges to be wrought."

Sturtevant in 1612, and Ravenzon in 1613, tried to

accomplish this wished-for process and failed, but in 1619, Dudley succeeded in making pig-iron with coal. He secured a patent, his bellows being worked by hand, his output but 7 tons per week. Under the Commonwealth, a patent was granted in which Oliver Cromwell was himself a partner. Dudley had estimated early in the century that there were 300 furnaces in blast for smelting iron ores with charcoal, with an annual output of about 180,000 tons ; and in 1620 he reckoned there were 500 forges and mills for refining and converting crude iron into malleable bars with an output of 75,000 tons. The residue was absorbed for cast-iron guns and mortars and ballast for ships. Harrison, speaking generally of our mines under Elizabeth, says, " They do bring forth so fine and good stuff as any that cometh beyond sea, and is of such toughness that it yieldeth to the making of clavicord wire in some places." On a progress in the north, Henry VIII. was able to see, within ten miles, no less than twenty-five coal mines and three forges for making iron. The ore was chiefly found in the woods between Belvoir and Willock in 1577-87, when the Earl of Rutland was busy reviving the old monkish industry of Rievaulx, and his iron was commanding the high price of £10 per ton. Labourers' wages in 1641 were twopence daily ; the smiths, forty-six shillings and eightpence per annum, with meat and drink. Farther north a coarse iron was produced at Bolton-le-Moors, and we learn that in the time of Henry VIII. they " burn sum canale but more these cole, of the wich the pittes be not far off." They also burned turf. At Horwich and Bury in Lancashire iron had been worked, but for lack of fuel, Leland informs us, " the blow shoppes decay there." Later on, the banks of the Derwent near Durham were lined all the way by mills, forges and furnaces for manufacturing iron and steel. Polydore Vergil says that the Southern Scots dig coal for fires, but possibly this applies to Newcastle. Little mention is otherwise made of ironworks in Scotland, where the celebrated Carron works were not started till 1760. Their carronades, 24 - lb. guns, had a great reputation, and added largely to their trade, giving employment to several hundred men. There was a guild of hammermen in Edinburgh from early times, but Froissart states that it was destitute of iron, even to shoe horses, when the French were there. Iron was largely imported in early days, for great quantities were intercepted in Scottish ships in the time of Henry IV. From it were made the broad round plates called gridles, on which the Scottish oaten cakes were baked. Dundee was once famous for armour, and when Defoe was there, the best bridles and horse-trappings were produced at Barnard Castle. In Ireland, Lord Cork established his ironworks near Youghal early in the 17th century, burning down the surrounding woods—a bad example, which helped to deforest that part of the country.

CHAP. II.—ROMANO-BRITISH IRONWORK.

R EADERS need hardly be reminded that iron-working was carried on by one or more of the numerous and varied tribes or races which occupied Britain in the Dark Ages, before the curtain of history was raised, through Cæsar's written accounts of his exploits in the invasion of this, until then, little-known isle. Antiquarian research has revealed that iron was known to us long before the dawn of history, in the dim past of the "Early Iron Age," as to which almost all our knowledge is derived from lake-dwellings and burials. The latter reveal the existence of widely differing races in varying stages of civilisation, and with distinct manners and customs. Cæsar landed in Kent on a coast occupied by Belgic Gauls. There were numerous other independent nations in Kent, Essex, and Middlesex, of whom he says little, while of those to the west he says still less. After a first British defeat, in which some of his forces participated as mercenaries, a powerful chief, Cassivellaunus, appeared in command of a fair, large-limbed race of warriors, distinguished by Cæsar as "Aborigines." These formidable enemies fought mounted and on foot, but most strange to him appeared the use made by them of two-horsed and two-wheeled chariots, which they employed in enormous numbers, and handled with great dexterity ; not, however, to fight from, but for rapidity of movement. Their tactics were "guerilla," and their occupation war and the chase, while they disdained fixed abodes, agriculture, and all the arts of peace. They dressed in skins and tattooed so extensively that their bodies assumed a bluish cast. Valour was, however, unavailing, for Cæsar crossed the Thames into their territories and captured their stockaded stronghold ; yet states that his foe, when totally defeated and a fugitive, was still accompanied by 4,000 charioteers. These people remained the secular and unconquerable foes of the Romans and Romanised Britons. Herodianus, relating the expedition of Severus, 250 years after Cæsar's invasion, observes that the Britons, the Greek equivalent for " Picts," the name given to them later by the Romans, were a most warlike and sanguinary race, carrying only a small shield and a spear, and a sword girded to their naked bodies. " Of a helmet or breastplate they knew not the use, esteeming them an impediment through the marshes." The conduct of Caractacus, one of their later chiefs, when prisoner in Rome, reveals that they were not devoid of noble traits. The walls of Severus and Antoninus were built to restrain their incursions, but they eventually suffered the fate of the Red Indians before advancing civilisation, and either died out, or perhaps left descendants in the Highlanders, whose manly qualities may be in some degree inherited from such noble ancestry. From Herodianus and later writers we learn that they encircled their necks and loins with iron rings, as an evidence of wealth, instead of gold, and went naked rather than conceal the tattoos of different animals which covered their bodies.

Their immediate connection with our subject is the immense demand for iron the use of chariots in such vast numbers must have occasioned. The wheels, three feet in diameter, were tyred with iron, and for these and their javelins and knives a large regular supply of iron was essential.

Iron in very large quantities must also have been required by the great fleet of 220 vessels owned by the Veneti, but partly manned by Britons, which were attacked and destroyed by Cæsar before he could attempt the invasion of Britain. They were built of oak planks bolted together by iron spikes as thick as a man's thumb, and the anchors were held by iron chains. Whether they were also beaked with iron like the Roman galleys is not clear, but they towered above them, and were far more seaworthy. Not being propelled by oars, they fell a prey, while at anchor and during a dead calm, to the Roman galleys. Such a fleet must in itself have required a large supply of iron, which no doubt came from Britain. The fact that it was exclusively employed between Kent and the mainland in carrying passengers and merchandise implies a large export and import trade, in which iron probably held a considerable place.

There can be no doubt that the iron industry increased vastly under the Romans, who managed to discover almost every deposit worth working. Few objects of iron of Roman date, as yet discovered in England, present special interest, being chiefly tools and weapons. Probably the few interesting iron objects illustrated in PART I. (Fig, 8—13) of the present Handbook belong to a time posterior to the departure of the Romans in 410 A.D., but no artistic use of this metal was made at that time, or previously, by the Romans in Britain. The inhabitants who had submitted must have been completely Romanised, and no doubt became accustomed to wear the helmet, greaves, and corslets, or other accoutrements of legionaries, in time of war. The residue of the Roman hosts and stragglers who remained in Britain were certainly of very mixed races. The subjects of the some eighteen British kingdoms were principally Kelts, and numbered Gauls, and no doubt Franks, Basques and Bretons, among them. During the two succeeding centuries these developed characteristics, embodied in the legends of King Arthur, wholly opposed to Roman ideas, when the serviceable qualities of armour and weapons were sacrificed to mere display. They were said to bear an evil reputation, in obsolete and fantastic panoply, as vain and fruitful in menaces, but slow and little to be feared in action. Their frightfully demoralised state, if not greatly overdrawn by Gildas, called for a day of reckoning. They presented a tempting and easy prey to the Frisian pirates, professors of rapine, who, as Teutons, were armed with the *fram*, or spear-like javelin, as described by Tacitus, and the gaudily painted buckler, but with iron *umbo* and rim, to which they had probably added a sword and dagger and some kind of simple head-piece. These prevailed, formed settlements, and dominated the country outside Scotland and Wales.

The next struggles on our soil were with mailed warriors, for the ravaging viking landed on our shores equipped in mail, the "war nets" of Beowulf, "woven by the smiths, hand-locked, and rivetted"; "shining over the waters," or in "the ranks of battle." Their shirts of mail, called "byrnies," attributed to the fourth or fifth centuries, are found in Danish peat-bogs, fashioned of rings welded and riveted in alternate rows neatly and skilfully by the hammer, if it be a fact that wire was not invented till near a thousand years later. This mail is distinctly represented on the Trajan Column, and wherever worn, whether by the Scythian, the Parthian, or Sassanian, its wearers successfully resisted even Roman discipline. Its weight made horsemanship imperative. The use of mail passed, probably by the amber route, from Western Asia to the Baltic, and necessitated weapons of extraordinary temper, especially the sword, invested with a mystic glamour which lasted to the age of chivalry. Famous swords are praised in the Sagas, and an actual specimen in the British Museum is inscribed in runes translated as "Awe inspirer." Swords and knives of this date are damas·cened and inlaid with silver and gold. The Norsemen harried the country by sea, or brought or obtained horses and raided inland, until settlements were granted them. From the Danes the exaltation of the sword passed to the English, and we find Ethelwulf, Alfred, and Athelstan bequeathing

B

their swords by will as most precious possessions, equivalent to a brother's or sister's portion. Thence it passed, in legend at least, to the Britons, King Arthur's sword Caliburn or Excalibur, ultimately presented by Richard I. to Tancred in Sicily, being no less famous than Arthur himself. The veneration for, or exaltation of, weapons and armour is thus traced to wearers of mail, and enormously added to the importance of the smith throughout these obscure ages, The umbos of the circular shields of the Anglo-Saxons are forgings requiring extraordinary skill. Germans and Danes had now fused into Englishmen a mixture of Angles, Frisians. Saxons, Danes and Norsemen, with whatever else had remained here after the Roman occupation. Such a mixture, perhaps hardly equalled in the present day, made for rapid progress, especially influencing our metalwork. Anglo-Saxon missals represent buildings with iron vanes and finials, doors hinged and covered with scrolled iron designs, all far in advance of any contemporary work on the mainland, and first developed in England. Several of these are illustrated in PART I. of the present Handbook. Meanwhile English customs had prevailed in war equipments, and the wearing of mail had fallen into disuse, or was at least optional or confined to the nobility. In the case of Harold, his mail armour was presented personally to him by Duke William. The Norman invasion, however, brought fresh hordes of mail-clad warriors, who prevailed ; and mailed cavalry thenceforward took the place of infantry until the disputes of the sons of the Conqueror led once more to English infantry taking the field. Thenceforth the Norman dismounted at the time of battle to fight on foot. Meanwhile the shirt of mail with short sleeves gradually developed by successive additions into the complete sheathing of mail which covered the man-at-arms from head to foot.

CHAP. III.—MEDIÆVAL IRONWORK.

WE may well believe that in the first decades following the Norman Conquest the smiths were chiefly absorbed in fashioning implements of war and appliances for defence. There could have been no lack of skilled workers, for British smiths of no mean capacity had been reinforced and amalgamated with those arrived from Saxony, Friesland, Denmark and Norway, and later by the armourers and *ferriers* of Normandy. A first care of the latest invaders was to build strongholds and fortify the towns, especially their entrances, which, because vulnerable, required an extensive use of iron to secure them. These strongholds were, if possible, moated, when a drawbridge of stout timbers securely fastened together by iron straps, and raised and lowered by iron levers, cranks, and chains, afforded the only entrance. First-class castles were defended by a gate-house on the further bank, and the main entrance through the walls was also flanked by towers with small window openings strongly barred, and a gangway closed by the portcullis, commonly of oak shod with iron, but sometimes, as reported of Pembroke and Newcastle by Leland, " *ex solido ferro.*" They were known as *cataracta*, since they might be dropped so suddenly as to entrap enemies endeavouring to force an entrance between these and the gates behind. Some castles have been described with even seven or eight of them, one behind the other, and the emplacements of six or seven are still visible in the main entrance to that of Devizes. Their great importance and extensive use in feudal times are commemorated in heraldry, for over thirty towns and many noble families bear the portcullis

B2

on their escutcheons, familiar later as the royal badge
of the Tudors.   Behind these were the gates of stout oak
requiring iron hinges, locks, bolts, and bars.   The portcullis
remains in a few castles, and there are even yet drawbridges to some of our moated dwellings.   The iron pivots
of the drawbridge to Stirling Castle, on a French principle
differing from ours, are yet *in situ.*

In addition to these were postern gates, either of wood
clamped and studded with iron, or of stout iron bars
crossing rectangularly, the horizontals being forged to give
passage to the verticals, or the converse.   Leland notices
exceedingly strong gates of this kind as still existing
at Tunstall and Raby Castles.   Warwick Castle possessed
an iron postern, and " Irongate " commemorates one
formerly at the Tower of London.   Such gates were
extensively used as defences to the smaller strongholds of
the Welsh and Scottish borders, known as yates, yats, or
yetts.   Forty - four of these still exist in Scotland.   They
are so fashioned that while the bars were being welded,
one half were passed through the verticals and the
remainder through the horizontals, very difficult forging,
but when completed impossible to wrench asunder.   These
were fitted to border dwellings and also to the border
churches until James I. ordered them to be discarded.   A
rarer form was the lid-gate, used in France, and swung on
central pivots, the name perhaps surviving in our " Lud-
gate."

Another use of iron practised from early times was to
sheathe the woodwork of doors, one such, indeed, being
discovered among Roman remains at Silchester.   Examples
may yet be seen at Durham and Maxstoke, the latter
strengthened by horizontal bars on which the Stafford
knot and arms appear, added, according to Dugdale,
in 1432.   The wooden entrance doors to Dunster Castle, in

Somerset, are stoutly latticed back and front with iron bars, probably not an unusual custom. Domestic buildings were secured by heavy bars pivoted to their gates, the ends passing into sockets, combined with bolts, locks, and chains, as still in use in some of the older colleges of our universities.

FIG. 1.—Method of working window-bars.

Windows in all important buildings throughout mediæval times were invariably secured by iron bars treated somewhat as the yetts. Chaucer refers in the " Knight's Tale " to " a window thikke of many a barre—Of yren gret, and square as any sparre." Various ways of manipulating these are shown in Fig. 1.

But it was especially to the windows of the treasure chambers of palaces and abbeys that most ingenuity was directed, for in these all portable worldly wealth was deposited. Extra strong trellises of heavy iron bars, threaded through each other in reversed directions, still exist in several of our cathedrals, one behind another. Such defences as the window of the old treasury in Canterbury Cathedral (Fig. 4) seem to have been efficacious, except in the historic robbery of Edward's treasure in 1303, kept in the vault beneath the Chapter House at Westminster. Here the windows are low and near the ground, but the original bars no longer exist. The treasures themselves were stored in trunks hollowed from trees and banded with iron, and fitted with lids hinged and locked ; or else fashioned from stout planks rudely adzed and similarly treated. In ancient days chests served all the purposes of safes, store-cupboards, libraries, wardrobes, cabinets ; and for travel. Some still preserved in churches date back at least to the 13th century, and may be earlier, and are so banded with iron straps in both directions that little of the oak is visible. They are always strongly hinged and secured by one or several locks, and at times also by padlocks. Frequently they are six or seven feet long, sometimes more, and furnished with handles, rings, and bars. Trunks usually have domed lids, perhaps a reminiscence of their origin, while in some early cases only portions of these lids are hinged. The jambs of the front and back of chests are often prolonged to stilt them from the ground. For travelling, coffers were made of lighter wood in pairs, easy to be slung on sumpter horses or mules.

From the 12th century a complete series of door-hinges exists, disclosing a remarkable diversity of design quite peculiar to this country, and a reflexion perhaps of the mixed derivation of its peoples. Happily the smiths seem to have been almost unfettered, while other craftsmen were somewhat constrained to follow the prevalent lines of architecture. Early church doors not only possess richly scrolled hinges, but the intervals are sometimes still filled with devices of religious import worked in iron. More rare are the even richer defensive linings to church doors, completely covered on the inside with scrolled or more or less geometric diapers. From their relative abundance, richness, variety, and early date, compared with corresponding work on the Continent, they confirm our claim that the fashion for richly worked iron in connection with ecclesiastical edifices originated in England ; which, in fact, maintained its pre-eminence in every kind of metal until the close of the fourteenth century. Well hammered, some gilt or tinned, if kept well oiled and cared for, especially when within the shelter of a porch or even a recess, they remained practically indestructible. A goodly series had been handed down, but unfortunately of late large numbers have disappeared through " restorations," though most old hinges can be kept serviceable by welding on new hooks and eyes, which are the essential parts.

Though so varied in detail, the decorative hinges associated with Norman architecture fall into two principal groups, the straps with simple or diverging scrolls, and straps proceeding from a crescent, the latter apparently preferred for its greater strength and resisting power. Most of the hinges of this period are bordered, hatched or diapered with chisel and punch marks, the most usual being a dot and lozenge design ; thus even the plainest straps, which usually taper, are rendered decorative. The pair from St. Albans Abbey, now in the Museum, No. 356–1889, belong to the group of strap hinges with diverging scrolls, of which they are perhaps the finest, and unique examples. They are peculiarly interesting from the design being clearly based on serpents combined with deeply serrated and primitive looking leaves.

Early examples of the crescent and strap form are illustrated in PART I., 3rd Edition, of the present Handbook, those from Stillingfleet (Fig. 26) and from Willingale Spain, Essex (Fig. 28), both with the serpents' heads ; and another from Eastwood Church (Fig. 29) in which this feature is lost. In another from Haddiscoe the crescents have become of almost rectangular form (Fig. 30). All are magnificent examples. Fig. 27 of PART I. presents an illustration of one of the diapered linings to doors. A still finer specimen is here illustrated from Skipwith Church, the basis of design being the intersecting circle of flat iron ; in this case it is bordered on both margins by roped edging, the fixing nails forming part of the enrichment. Each interspace is occupied by a cross, circle, or fylfot (Fig. 5).

Doors and grilles of bronze were not unknown in the Roman Empire, and were of scale, trellis, and lattice design ; and no doubt such existed in England in early days. They were re-introduced, but made of iron, to safeguard the choirs, chapels, and chantries of cathedrals and abbeys, in which valuable reliquaries and shrines were shown to pilgrims. They must have been a necessity in Anglo-Saxon days, and the smiths who covered doors with intricate scroll-work no doubt designed their grilles in a similar spirit. None could have been more admirably conceived for the purpose than those beautiful existing screens of scrollwork, strong and unclimbable, through which nothing could be

purloined. Quite transparent and safeguarding, they appear
rather veils than screens. Every one of our great monastic
buildings, cathedrals and abbeys must have possessed
such in some form, yet but four now remain, two being
mere fragments. Portions of the great grille of St. Swithin's
shrine still exist, the position where it was fixed being even
now easily traced. It excluded pilgrims from the choir, south
transept and nave of Winchester Cathedral ; but its remains
are now but a patchwork against a door in the nave. The
design is based on C-shaped scrolls, of lengths permitting
them to be arranged in bundles of three, one within the
other, and strapped back to back by iron ties, the ultimate
volute of every scroll being bent and secured to form an
open-work cinqfoil. Six of these bundles, secured by ties,
form a panel in a separate frame. Two such panels exist,
and there are two others of later date in which the scrolls
are more compactly welded together with ends bent in
trefoils (see the reproduction in the Museum, No. 1891–27).
A similar design occurs in France at Le Puy, assigned to the
twelfth century, and there are others in Spain to the north
of Madrid which must be somewhat later. Our Winchester
example dates from about 1093, in the reign of William
Rufus (Fig 6). Thus it is more than probable that the design
originated in England. A perfect grille, comprising a gate
and four panels, closes St. Anselm's Chapel in Canterbury
Cathedral, completed between 1093 and 1109 (Fig. 7) ; it is of
C-scrolls recurved slightly at the ends in form of shepherds'
crooks, and, except in one panel, tied in couples placed
vertically. Our only complete choir grille, dating from the
second half of the twelfth century, is also fashioned of plain
vertical C-scrolls tied together in stoutly framed panels :
a second horizontal at the base framing a narrow band of
pierced quatrefoils. Unfortunately, the original cresting of
spikes has been removed to make way for a row of

gas-burners. This is in Lincoln Cathedral (PART I., 3rd
Ed., Fig. 40). Grilles of the same design are found
erected by the Crusaders in Jerusalem, and there are
several in France. An iron grating formerly shut off
the high altar of Canterbury Cathedral. As Becket
approached it on the day of his death, the gate was
open, and he forbade it to be shut. Later, iron grilles
protected his shrine. The iron grating put by Abbot William
in St. Albans when he removed the saint's shrine to
the middle of the church in 1275, was probably of this
design of C-scrolls ; if so, the small portion still preserved in
the Abbey may be a relic of it.

In the reign of King John we hear for the first time of
a bridge protected by an iron gate at Stratford-le-Bow,
erected by the collector of tolls, till the Abbot of Waltham
broke it down, as it hindered his waggons.

When we pass to other work of the mediæval smith for
churches, the lover of ancient craft-work must mourn.
Where are the great chandeliers, candelabra, standards, and
branches for lights ; the lecterns, chairs, font-cranes, and
screens which formed part of the furniture of every important church ?

The only example surviving of a portable lectern, of a
type far from uncommon on the Continent, is in Chippenham
Church, Cambridgeshire. The sloping desks, back and
front, are supported on a plain iron stem and tripod foot,
and comprise four bars intersecting like the Union Jack,
three fleurs-de-lis forming a cresting.

Henry III. was the first extensive patron recorded of
English smiths and craftsmen generally. His indents for
ironwork were unusually large, and comprise nails in immense quantities, window-bars, finials, iron lattices, locks,
an iron trellis for staircase, iron " kevils " with chains to
shut the glass windows, lecterns, and branches for candles.

He also fixed an iron grating round the shrine of St. Amphi-
balus, which he later removed to a more central position in
St. Albans Abbey. The fashion of the grillework now
completely changed with that of the architecture, becoming
geometric. In Salisbury Cathedral are the monuments of
Bishop Brougham, 1246, and Bishop William of York, 1256,
one on each side of the choir. The arches beneath which these
repose are filled with gratings of intersecting vertical and
horizontal bars, forming panels fourteen inches square, the
iron one inch on the face and five-eighths thick. In each
square is a pointed quatrefoil of thinner iron, the inter-
spaces forming a secondary pattern of lozenges which read
diagonally, thus forming a pleasing diaper. Gough shows
an identical grille to the tomb of Bishop Bingham, who died
in 1247, placed beneath a cusped arch. A grille of precisely
the same design seems to have been destroyed in the Priory
Church of Christchurch, Hants, where part of the work has
since been utilised to form a low gate to the porch. A later
grille in Canterbury Cathedral of geometric diaper, of the
time of Edward I., 1303, closes the entry to the choir from
the nave, and is fashioned in the manner of Oriental lattice-
work. A portion is reproduced in the Museum (No. M. 1912-2)
(Fig. 8).

Henry III.'s gifts to Westminster Abbey, then rebuilding, are the most sumptuous on record, including a
shrine of purest gold for the relics of St. Edward, and a
great crown of silver to set wax candles on, and three silver
basins for lamps. His goldsmiths were Odo of London,
and his son, Edward Fitz-Odo, or Edward of Westminster ;
and Pietro Cavallini, brought from Rome to do the marble-
work and mosaics, who, before returning, prepared, no
doubt, the tombs of Henry, his son Edward, and of Edward's
wife, Eleanor of Castile. The beautiful effigies of Henry
and Eleanor, of gilt bronze, are by William Torel of London.

These superb tombs fill the arches of the north side, which it was necessary to complete and protect. The Royal smith, Master Henry of Lewes, whose pay was 6d. per day, while other smiths had but 3d. or 4d., made the grille for Henry's tomb in 1258–59. It was of somewhat severe design, but silvered and gilt. He also, no doubt, made the eight small moulded and gilded quatrefoil flowers securing the large porphyry slabs, which may well have suggested the method of stamping so soon to follow. Henry of Lewes was also paid in 1290 for the grille to Edward the First's monument, seen, as illustrated by Dart, to have been severely plain, of twelve vertical bars with six horizontals, blocked out at the intersections as in all the contemporary window grilles. The verticals are five feet high, and appear from the front to terminate in large fleurs-de-lis, but were actually carried over the tomb to form a baldequin, although there was probably no effigy. In addition, there were three standards surmounted by human heads representing elderly bearded men with long visages, thus resembling the monarch. Edward presented, perhaps at much the same time, though only coming into the accounts in 1295, three iron chairs adorned with human heads for the use of the precentors in St. Paul's Cathedral. These were plated with silver and gilt, and no doubt the tomb-rail was similarly treated. The iron grille to the monument of the King's brother, Edmund Crouchback, who died in 1296, is shown of identical design, but without busts. These rails, with all the rest of the ironwork, were swept away in 1821 by an egregious " Committee of taste," and only the Eleanor and the Henry V. grilles have since been discovered and restored, no doubt owing to the " mediæval revival " in 1850.

The smith Henry of Lewes first appears on the Palace Rolls in 1253. He died in 1291, leaving houses in both Lewes and London to his daughters and wife.

He apparently made no use in these works of stamping in relief by aid of chilled iron dies, which gives such richness to the later Eleanor grille. Professor Lethaby, however, suggests that, as a Sussex man in Court favour, he would probably have been employed on the grilles in Chichester Cathedral, which filled six arches of the east end of the Choir, each with a different pattern of scrollwork. In one only of these the scrolls end in discs with fleurs-de-lis and roses stamped in relief, the raised peripheries continued back for a short distance along the scroll as a mid-rib. The rest are variously scrolled without the use of stamps or modelling. Thus one of these grilles would appear to present a very early example of the use, by the smith, of stamps from prepared dies. The grilles are also of interest as the fourth and latest of the rare examples of iron screens fashioned of scrollwork remaining in England (PART I., 3rd Ed. Fig. 41).

The grilles had until 1860 been hidden by galleries, and were then discarded as too dilapidated for use. Fragments were rescued and put together by a private resident for the Cathedral, but again discarded, when they found their way to the Museum. Standards with fine terminals of stamped foliage have also disappeared, as well as part of a richly scrolled grille, tool-marked and with stamped rosettes and heads.* The treasury chest with five locks, in the Consistory of the Cathedral, is no doubt by the same smith, and measures eight feet in length and twenty inches in height, with vertical straps strongly ribbed and terminating in stamped leaves and rosettes. It bears two ring-handles on each end, and was known in the days of Henry VIII. as the Vestry Coffer " where joyelles lyeth." Professor Lethaby thinks that Henry of Lewes, as the King's smith, may have executed it, and also the rich linings to the great

* *Architect*, June 19, 1896.

doors of Henry III.'s Chapel at Windsor, now closing the
east end of St. George's Chapel (PART I., 3rd Ed., Fig. 48).
The design for each of these doors is of three large vesicas, the lower incomplete ; their interspaces entirely filled
with boldly designed scrolls, branching in pairs from the
central stem, ending in rosettes, and with trefoil leaves
both developed and in profile springing from them on
either side and together forming a rich but irregular
diaper entirely covering the doors. The leaves and
rosettes were stamped in relief, while hot, by the use
of chilled iron dies, the process apparently invented by
Henry, while the stems and scrollwork are deeply grooved.
Seven or eight separate stamps are used for the leaves and
six for the rosettes, of which five are circular, while another,
of heater shape, is used but once. The latter and two other
of the dies are found on the left-hand door only. All are
intended to represent flowers. In the apex of each of the
two lower vesicas on each door is, in addition, a larger
stamp, of vesica form, with " *Gilebertus* " in raised letters
and a small cross, referring, as Professor Lethaby suggests,
to Gilbert de Tile, bailiff of Windsor in 1256. Small dragons'
heads hammered in full relief occur in some of the inter-
stices of the ornament. The Chapel was commenced in
1240 and completed in 1256, and the ironwork remains
still in perfect condition, and unrestored.

The hinges of Merton College, Oxford, owing to their
crescentic form, have been assigned to this date, and some
of the Windsor stamps are apparently used, these corresponding also in other details to Henry of Lewes' work
(Fig. 9). The same characteristics, with additional stamps,
rendered necessary by the smaller scale and more delicate
nature of the scrollwork, recur, applied to the aumbry
doors, in the vestry of Chester Cathedral,* and to the lid

* See the plaster reproduction in the Museum and illustration in
PART I., 3rd Ed., Fig. 49.

of one of the two cope chests in York Minster. The
fine work lining the Chapter House door of York
Minster is also by the same hand, and in both the small
dragons' heads are found amidst the scrollwork.

Henry of Lewes was succeeded in the office of smith to
Edward I. by Thomas de Leighton, who was in 1294 employed
on the Eleanor grille in Westminster Abbey. In this work
foliage produced in high relief, by the use of chilled iron
dies, reached its fullest development in this country. Such
work hitherto applied to the wood surfaces of doors and
chests is here, for the first time, applied to an open framework of iron, arched forward and surmounted by a row of
iron tridents, the function of the grille being less perhaps
to safeguard the finely gilt effigy of Eleanor of Castile on
her tomb than to secure the immensely valuable shrine of
Edward the Confessor behind it. The grille consists of eleven
distinct panels of conventionally treated stems and foliage in
relief, none being exactly alike, while two markedly differ in
treatment. The cost was £13, equal to about £118 of our money
(Fig. 10). A reproduction is in the Museum, No. 1888–425.
The similarity of the work, which comprises dragons' heads,
implies that Thomas had been associated with his pre-
decessor in the King's work. The tridents are of different
fashions, somewhat like those over the chantry screen
by William of Wykeham at Winchester, which, however,
are not earlier than 1367. Another trident with spikes
is over the door to the Chapel of St. John in Westminster
Abbey. Several chests are also of his work, as well as
hinges at Leighton Buzzard, whence he probably came,
Eaton Bray and Turvey in Bedfordshire, and Tunstall,
Colchester and Norwich. John le Fleming made an iron
door to the new treasury at the Tower for Edward I.,
after the great robbery at Westminster in 1303, for £3 17s. 4½d.,
but none of his work has been identified. The Museum

possesses an interesting cupboard decorated with iron, scrollwork, the ends of which are slightly stamped: it is provincial work, probably of the fourteenth century, and is said to have belonged to the last Abbot of Whalley, Lancashire (Fig. 11).

Until the thirteenth century, monuments, even of kings and bishops, were little more than effigies in bas-relief, and not raised much above the pavement level, as those in the Temple Church; but in the reign of Henry III. tomb-like sepulchres came in, often with life-like effigies lying in state upon them. Such stately objects, if of benefactors, were usually placed between the choir and the ambulatory instead of grilles, which they no doubt have often displaced. The exceptional value of the gifts to the shrine of Edward the Confessor at Westminster led to the tombs on the north side being protected by iron grilles, though raised much above the ambulatory. No tombs were erected on the south side until the reign of Edward III., when the three spaces were occupied by monuments to himself, his queen, and his grandson Richard II. There were probably protecting rails before these tombs were set up, retained to protect the shrine, but if so they were of vertical bars simply treated. All were swept away early in the nineteenth century.

It does not seem, however, that tombs were necessarily considered to need the protection of railings till effigies were carved in alabaster. The earliest recorded as protected by an iron grate is the very costly marble effigy of Marie de Coucy, made in Bruges for Newbattle Abbey in 1214. The marble effigy of Queen Philippa, made for Westminster Abbey by Hawkin of Liége in 1367, appears to have been the first in England protected by a railing of vertical bars. These supported a heavy moulded cornice and battlements with massive standards at the angles, also battlemented,

and with prickets. The work was probably by Peter Bromley, smith to the Abbey, who retired two years later with a
pension of 100s. per annum for good behaviour. The
solitary railing, only spared to the Abbey because it happened to shut off the ambulatory from St. Benedict's Chapel,
is also of vertical bars set diagonally with spikes seven inches
long above the moulded cornice : it comprises five massive
standards with moulded buttresses and battlemented caps,
and protects the tomb of Archbishop Langham, erected
three years after his death by grateful monks in 1376. The
gates to the physic garden at Oxford are of nearly identical
design and workmanship, and have probably been removed
from one of the chapels. The rail of vertical bars at Lincoln
to the monument of Catherine Swinford, John of Gaunt's
third wife, who ·died 1403, differs only in the reappear-
ance of the battlemented cornice and increased height of the
battlemented standards, which bear prickets. The tomb of
the Black Prince at Canterbury needed no protection, being
of bronze, and its railings were not added until after the
alabaster effigies of Henry IV. and Joan of Navarre were in
position in 1437. In 1440 Archbishop Chicheley erected a
monument to himself, although he survived till 1443. All
these tombs at Canterbury have identical railings, by the
same smith, of plain vertical bars set diagonally under a
battlemented cornice bearing stamped lions' heads and
fleurs-de-lis, and six battlemented standards with buttress
supports; those to the Black Prince having prickets, apparently a mark of extra dignity. They are without spikes,
an inconvenience realised as early as in 1443 when a German
thief took refuge inside one of them from his pursuers until
forcibly extracted by the citizens. These railings may all
have been contributed by the Archbishop, and their form
seems to have been prevalent from at least 1367 to 1440. Not
being considered of decorative value, the removal of such

c

railings from churches has in later times been almost univer-
sal. Such severely plain railings at last gave place to richer
treatments. The railing to the Fitz-Alan tomb at Arundel,
1415, is still of plain vertical bars, but set diagonally below

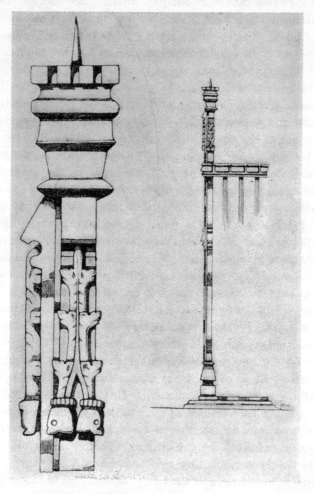

FIG. 2.—Railing to Fitz-Alan tomb, Arundel Castle. 1415.

a finely moulded and battlemented cornice, from which the spikes have recently been removed. The decoration is concentrated in the ten moulded and massively buttressed standards, on moulded bases, which rise sixteen inches above the cornice, and finish in moulded and battlemented caps bearing prickets. On the outer faces of these standards are carved narrow crocketed arches springing from grotesque heads (Fig. 2). Another departure is seen in the rail to Archbishop Chicheley's monument, 1440, in Canterbury Cathedral, in which the spikes, rising above a moulded cornice, are forged into stellate heads like snow crystals, alternating with smaller trefoil spikes; the standards being massive and lofty with octagonal moulded caps and prickets. Such departures probably influenced the design of the rails to both Sir Thomas Hungerford's monument in the chapel at Farleigh, Somerset, and that to Bishop Beckington's chantry at Wells (PART I., 3rd Ed., Fig. 62). Sir Thomas fell at Agincourt, and his monument is probably anterior to that of the Bishop, who died in 1464 (PART I., 3rd Ed., Fig. 61). The important innovations seen in these are, perhaps, partly due to the railing to Bishop Fleming's monument at Lincoln, about 1431, in which the plain verticals are held between wide bands with slightly raised mouldings pierced with reversing trefoils within vandykes.

An almost unique instance in England of a grille composed of open quatrefoils formerly formed part of an elaborate rood screen in Chichester Cathedral, erected by Bishop Arundel between 1459 and 1483, probably in 1478. It formed two gates of three panels, each of sixty-four quatrefoils of delicate work, framed in five-inch squares. These gates were " re-fashioned " in the restoration, when some of the panels found their way to the Museum (No. 592—1896, Fig. 12); the remainder now form gates in the grille of

C2

the Lady Chapel. There were formerly two earlier oblong panels of finely welded quatrefoils in contact, collared together as in Italian work, but they were much decayed, thirty years ago, and, being banished to a rather damp crypt, seem no longer to exist.

More ambitious monuments are the chantries erected by famous personages soon after this date. The very fine one to Duke Humphrey of Gloucester in St. Albans Abbey may be somewhat later than his death in 1446. The iron grille is on the ambulatory side, eight feet high and some fifteen feet in length, and consists of a framework of massively moulded vertical and horizontal bars forming forty-two oblong panels in three tiers. The general scheme is a chequer of trellis in a stout frame, all painted blue. All the intersections are riveted, the heads of the rivets hammered and filed into quatrefoils and cross-hatched and gilded. Over the heavily moulded cornice is a border, pierced in quatrefoils, with a central cross. A portion is reproduced in the Museum (No. M. 1914–15) (Fig. 13). In Sandford's view of the chantry a bold cresting of large crosses and small trefoils is seen over the cornice.

Chantries with vertical bars are of the same date. Lord Hungerford's, in Salisbury Cathedral, dates either from 1429 or 1449, and stood in the north arcade of the nave, each front forming three vertical panels between massive standards painted heraldically, in blue, gold, vermilion, etc. It was removed to the choir by Lord Radnor in 1779. The chantry of Abbot Wheathampstead, at St. Albans, is somewhat similar, though probably not erected until his death in 1464. It is divided into three vertical panels by four massive buttressed standards, beneath a heavy coved and battlemented cornice nine inches deep. A shield with the Abbot's rebus is fixed to the cornice over each standard. Far more ambitious is the famous monument of Henry V. in Westminster

Abbey, the grille made by Roger Johnson, between 1425 and 1431, when, owing to delay, he was empowered to arrest and press smiths to complete it. Standing within an arch between stone turrets, with stairs leading to an upper chamber, to which Henry had presumed to remove the altar of St. Edward, the front forms a large central panel with a gate on either side, the whole of twenty-four narrow vertical panels in two tiers, and in heavily moulded frames, rounded above, and each comprising five quatrefoils below a trefoil. These are constructed of short bars punched through the centre to form cuspings, arranged to cross diagonally between others set vertically and horizontally; resulting in quatrefoils recalling Eastern lattice work. They are backed by slightly overlapping pierced sheet-iron plates, giving richness and depth: the effect may be realised from a part reproduced in the Museum, No. 1888—426. The upper tier is separated from the lower by a band with raised margins, which originally bore swans and antelopes in relief. It finishes above in a wide battlemented cornice. The arch above it is closed by a grille of purely architectural design in heavy half-round iron, backed by pierced sheet iron, suggesting the coming fan tracery. A narrow border of delicately worked cusped quatrefoils forms a margin to the whole (Fig. 14). The back of the chantry was formerly closed by vertical spiked bars and cornice. These are now removed. The whole was richly decorated in gold, vermilion, and blue, the cornice emblazoned heraldically with groups of three lions alternating with three fleurs-de-lis. Beneath reposed the wood effigy of the great warrior encased in silver plate, the head of cast and chased silver; while above was the altar of St. Edward with its valuable gold plate. The handle of the door has escutcheons pierced with trefoils.

The screen at Arundel may be a few years earlier, but

still of the time of Henry V. It consists, beneath a bold
and heavily battlemented cornice, of a large pair of gates
and two fixed panels, constructed of twenty-two vertical
bars, two buttressed standards and a frame, crossed by
four horizontals, forming five rows of narrow vertical panels,
containing small cusped and pointed arches, the apex
carried up in a long spike through the horizontal into the
arch above. This simple expedient unites the whole strongly
in a pleasing manner. On the cornice above the arches
stamped lions' heads alternate with roses. A panel is
reproduced in the Museum (No. M. 1911–8) and here
illustrated (Fig. 15).

Frames for the pall, known as herses, and to hold
candles during the lying-in-state, were temporarily erected,
but occasionally formed a permanent feature of the monument, as in that to Mary of Burgundy, which was of forged
iron, designed by Jean d'Houry. The arched bars over the
monument to Edward I. in Westminster Abbey, which was
without permanent effigy, were probably to support a rich
pall. The fine effigy of the Earl of Warwick in the Beauchamp Chapel, Warwick, 1439, was hooped over with copper-
gilt staves. The effigy of Duke Robert of Normandy in
Gloucester Cathedral is similarly treated, but in iron. Miss
Fiennes noticed this iron grate, but by whom placed there
is now unknown. It has been credited to Henry VII., and
to Cromwell, whose soldiers are said to have partly destroyed
the wooden effigy. A herse was formerly fixed over the
monument of Lady Fitz-Alan, Countess of Arundel, in the
church there, the sockets and some fragments of iron still
remaining. Twisted iron rods bent into arches and holding
prickets stretch over a Marmion tomb in Tanfield Church,
Yorks. A rare example in the Museum is from Snarford
Church, Lincolnshire. It consists of a low rail of vertical bars,
alternately plain and twisted, supporting a long brass plate

between iron bars with an engraved inscription in black-letter, and a similar plate below of sheet-iron bearing semi-circles in relief, arranged in pairs, the lozenge spaces between being pierced in pointed quatrefoils and vesicas. Above is a cresting of inverted arches finishing in tufts of lily flower and leaves, bearing the prickets, and alternating with fleurs-de-lis. Within the spandrels are pointed trefoils of iron, and at either end are elevated clusters of indented leaves— one in the centre erect, and the two at the sides drooping. It seems to have been supported by four straps with applied spindle ornament attached to the masonry, but it is obviously part of an important structure, perhaps dismantled during the Reformation, when objects of iron, having no intrinsic value and being incombustible, were spared. It may date from the end of the fifteenth century, when the St. Paul family were lords of Snarford (Fig. 16).

The rail which closed Dean Urswick's chapel in St. George's Chapel, Windsor, a little later in date, has been mutilated and removed to a new and less happy position. It finishes above in a moulded cornice with a fret of trefoils from which rise groups of spikes, a centre upright, and four drooping, but all barbed vertically, so that each group may serve as a cluster of prickets for five tapers. Below are six standards buttressed and with shields of arms, now rearranged against the wall, with only three vertical bars between the standards, the whole resting on two horizontal bars for base. The date is about 1500 to 1520. Facing the old Urswick Chapel is that of Charles Somerset, Earl of Worcester, who died in 1526. The enclosing rail is now of brass, but above it is the older cresting of similarly barbed spikes for tapers, varied by five fleurs-de-lis of plate iron. Within, a fine contemporary railing of brass surrounds the altar tomb and effigy.

Mediæval window-grilles, though seemingly much alike,

differ considerably. Apart from those of Canterbury Cathedral, following the lines of the glazing in the French fashion, the vertical bars may be round or square, the latter frequently set diagonally. The bars at Faringdon, Berks, show the horizontals with welded blocks, and those at Orston are bent over and punched for the vertical to pass. Verticals often end in barbed arrow points, spear points, or may be forked. A more architecturally designed grille in the Beauchamp Chapel, Warwick, is of round bars hammered and riveted where they intersect ; another at Hereford is of flattened iron, worked into cusped and intersecting arches, dating from about 1460. Until the fine tomb of Henry V. was erected, in which even the stonework is said to have been gilded, English monarchs had been content with the altar tomb and effigy not materially differing from those of nobles and prelates. Henceforth these no longer sufficed, and the regal splendour of the resting-place of Henry V. was, in turn, eclipsed by the marvellous gilded iron gates and piers for the monument of Edward IV. at Windsor (Fig. 17). Beside this all previous works in iron sink into insignificance. Despite the suggestion of foreign workmanship, which had deceived Gough, Pugin, Wyatt, Burgess, and, indeed, all other experts, it is now established on documentary evidence to be the work of Master John Tresilian. This " chief smith at London " is first heard of in 1478, when looking after the making of a great anvil, getting carvings probably for models, and so on, and seeing them safely to Windsor, at a cost of £20 6s. 8d. His task at Windsor took six years to accomplish, and he was paid at the then high rate of 16d. per day, perhaps including assistance. The work for the actual monument comprises the pair of gates, 5 feet 6 inches high, between two half-hexagonal piers surmounted by eight open cressets or lanterns, nine feet high in all. The design, purely architectural, is Perpendicular verging on flamboyant, in the

extreme of elaboration and richness, and represents a
storeyed building with buttresses and oriel tracery windows.
It is carried through in a series of two-light windows, under
very rich overhanging canopies, between crocketed and
pinnacled buttresses, rising to support four even more richly
wrought hexagonal lanterns or cressets. That details were
left to the smith is apparent from the vastly different scales
in which identical architectural details are rendered. Some
traceried windows of the same design are twice, and even
four times, the scale of others. All the pieces are carefully
filed, riffled, chased and fitted with marvellous precision.
Miss Fiennes some 250 years ago remarked in her diary :
" There is in the Church a tomb and vault of the Duke of
Norfolk's family with steele carvings all about it, very curious, and, to add to its variety, it may be all taken apart,
piece by piece, and put up in a box. It 's a very large thing
and great variety of work, and this on the right side of the
altar." (Diary, p. 285.) It is, in fact, so constructed of
separate pieces fastened by cotter pins that it may readily
be dismantled. The earliest reference to it is in the MS. of
about 1610, with a sketch and note that Edward IV. was
buried in this place. Sandford gives a good engraving of
it in its original place, " the north arch near to the high
altar," where it then was between two piers in the north
ambulatory of the choir, where the original fixings still
remain. It was, in any case, only intended to be seen from
the front, and must have formed part of some important
monument since destroyed. He continues : " The King lies
in the new chapel whose foundations he himself had laid
under a monument of steel polished and gilt, representing
a pair of gates between two towers, all curious and transparent workmanship." Professor Lethaby regards it as an
adaptation of the gilt stone front to Henry the Fifth's
chantry ; while Sir William St. John Hope thought the

design borrowed from the stall canopies close by. Mr.
Aymer Vallance calls it purely English. As workmanship,
it is far more characteristic of Flemish or French work, and
it is not obvious that anything previously produced in
England can have directly led up to it. In any case we
must now be far more careful before assigning similar workmanship to a foreign source. Nothing further is known of
Tresilian, though it is obvious that the locks and other
furniture of the chapel are by the same hand. The most
important piece is the large octagonal alms-box, supported
on stout octagonal bars with moulded caps and bases, precisely following the lines of the stone shafts immediately
behind, thus clearly indicating that it was designed and
probably produced on the spot. The eight facets of the box
are four times higher than wide, each finishing above in an
applied arch on buttresses, with high crocketed finials and
the letter ℏ on the panel between. The cover was originally
flat and moulded, the whole structure held together by
cotter pins. A decorative summit of four bronze turrets
with slots is, no doubt, a later Tudor addition. Remarkable
also is the *guichet* of the door leading to the Royal pew ;
oblong with sloping top and front projecting about one and
a-half inches, of rich flamboyant tracery on brackets, attached
to a plate, between buttresses with crocketed finials, and
equally rich tracery above and below. On the reverse side
of this door is a hinged shutter with bolt, a bent closing-
handle, and the large bolts necessary for securing the door.
The handle-plate is circular, seven inches in diameter, with an
extremely rich border surrounding the whole, within which
is a garter in relief enclosing a pierced rose window of intersecting tracery. The lock-plate, six inches square with
moulded edge, is also of rich vesica tracery, completing the
furniture of this very important door. Two locks, each
about one foot long and five inches high, to the north and

south entrances to the choir, are also noteworthy. They differ in detail, but both are divided into three panels of rich tracery, framed by borders of half twists with beaded ridges. There are also several good bar handles and locks deserving attention, especially the lock to St. John's Chapel, 1522, and that to Sir Reginald Bray's Chapel, in which the " bray " for flax-combing appears in relief on a shield (Fig. 19). Repeats of these, also in iron, some twenty in all, are placed variously on the stone and woodwork. Several of the window-grilles to the chapel are worth notice.

At Eton there is little Tudor or earlier ironwork visible beyond the remains of one or two handle plates in the cloister. King's College, Cambridge, is, however, well provided with fine locks and handles of similar design and date, and quite possibly by Tresilian. The locks to the chapels on the north side of the nave and entrance to the choir deserve especial attention, resembling those in the choir at Windsor, but more perpendicular and less flamboyant in feeling, and of somewhat less elaborate and delicate work. They may antedate the Windsor examples. There are also a few of interest on the south side of the nave. The curious memento left by Edmund Audley at Hereford, where he was Bishop, on his translation to Salisbury in 1492, is of much the same period (Fig. 18).

Smithwork had now reached its zenith, and further progress seems arrested. The perfect and relatively simple pair of gates closing Bishop Alcock's Chapel in Ely Cathedral date from about 1488, and open under a Tudor arch. They are of plain vertical bars, finished above with small cusped arches between, and lower down shorter intermediate bars with large and beautifully moulded fleur-de-lis heads just clearing the upper horizontal, while slightly below is a fine lock with pinnacles and cresting representing a gateway ; and at the base a second strong moulded horizontal pierced with quatrefoils between the vertical bars.   They are still serviceable as when first erected.   In a somewhat earlier rail, about 1480, in Thame Church, the verticals end above in moulded spikes and strong welded fleurs-de-lis.   The rail to Dr. Ashton's tomb in St. John's Chapel, Cambridge, is of early sixteenth-century date, of vertical bars set diagonally beneath a cornice and foliated scrolls bearing an inscription.   The six standards are massive and buttressed, carried up above the cornice in angulated twists and moulded cap, supporting the tun and ash of the doctor's rebus. Henry VII. left instructions for a sumptuous monument and grille, the latter made in London, during his life, of gilt bronze, at a cost of £1,000.   Perhaps no smiths were capable of carrying it out, for thenceforth smithing under the Tudors ceased to be a fine art.   Foreigners were preferred and greatly encouraged to settle.   Nicander Nucius reports that in London there were strangers from many of the nations of Europe, especially with regard to the arts connected with the working of iron.   Henry VIII. showed no concern in the interests of English smiths or ironworkers, and patronised, yet alternately threatened, the German Steelyard, in whose hands most of the iron trade was then concentrated, and from whom he extracted large sums of money.   The Flemings were otherwise perhaps the most favoured intruders.   The chief smith for making or supplying the builders' ironmongery, such as window-frames, stays, hinges, and grilles during the building of Hampton Court was John of Guylders.   Immense quantities of iron nails, locks, chests, etc., came from the Dutch, no doubt originally from Germany.

The railing now restored to the beautiful altar-tomb of Margaret Beaufort, Countess of Richmond, in Westminster Abbey, was completed in 1512, and is described in the report to the Lord Treasurer Burleigh as " a closur of yron about her with vaynes & armes," and remained in perfect condition until sold with the rest for old iron in 1821. It is of plain vertical bars below the cornice, with javelin points fixed over it, alternately converted into fleurs-de-lis by a delicately forged scroll on either side. The moulded horizontals with rope edges were formerly enriched with applied rose and portcullis alternating, heraldically treated and gilt. The six standards are buttressed, and surmounted by a twisted staff formerly bearing fleur-de-lis edged banners and finials. The work was recently reclaimed, and after necessary renewals, presented and refixed in the Abbey, but shorn of its heraldry, the most interesting and important feature. The only remaining rail in Hereford Cathedral, where there were forty-six tombs, is to Bishop Booth's monument, 1535. Another, in Lincoln Cathedral, is to the tomb of Bishop Longland, confessor to Henry VIII., 1547. The finely worked gates to Bishop West's Chapel in Ely Cathedral were made in 1515, though the Bishop, who had lost the King's favour through opposing his divorce, lived on to 1533. The gates open under a Tudor arch, their tops filled with intertwining stems in the manner of the Antwerp well-cover, bare of foliage, but with exaggerated spines and dwarfed roses. Below are twisted verticals supporting eight narrow cusped ogee arches, the tracery centring in a rose, with a shield and fleur-de-lis. A wide band forms the lock-rail, with an ornament of conventionally intertwined stems and thorns, the lower half of plain bars finishing at the base in a double horizontal with plate tracery between. A Renaissance touch is given by classic balusters in semi-relief covering the meeting bars (Fig. 20). The railing to the monument in Salisbury Cathedral to Henry VIII.'s sister, the Queen of France, and later wife to Charles Brandon, must have been an interesting one, since it attracted the notice of Miss Fiennes in her book of travels. It was, however, destroyed, with that of Lord Gage's tomb, in a " restoration." The story of the development of iron rails to monuments is sadly incomplete, owing to wholesale destruction, continued uninterruptedly from Cromwellian days almost to the close of the nineteenth century. Of some forty existing in Westminster Abbey till 1821, only three are reinstated. The Great Fire destroyed, perhaps, as many more in St. Paul's, and unknown quantities in City churches, for ten in Christ Church alone were railed, while other abbeys and cathedrals suffered as severely. Thus Exeter, Chichester, Worcester, Tewkesbury, Shrewsbury, Bristol, Coventry, Gloucester, Salisbury, Cirencester, to mention only a few in the West Country, are either totally devoid of ironwork to monuments, or contain a stray specimen only, saved because it happened to fulfil some utilitarian purpose.

In those days iron railings were rarely used in the open, though trellises like modern hurdles sometimes appear in mediæval illustrations of gardens. One exception is frequently noticed, in memoirs, from the Duke de Najera's in 1543, to Camden, Defoe, and even later writers: a dwarf iron railing, either for appearance or safety, had been fixed on the stone parapets of the long bridge of eleven arches over the Medway at Rochester, at the sole cost of Archbishop Warham. Its novelty must have been its sole attraction, for an excellent print shows it to have been utilitarian, of merely plain spiked bars let into the stone and held together above by a single horizontal.

Window-bars remained in request throughout Henry's reign, the accounts of Hampton Court showing that bars, by his smith John of Guylders, were needed in quantities. The verticals were termed " standards," and the horizontals " stay-bars," while insertions welded in for the passage of one bar through another were called " locketts." Little, if any, original Tudor ironwork exists, however, in this superbly historic building. The use of constructive iron in

building seems to have been introduced to England by
Henry VIII., for the aerial look of Nonsuch is largely attri-
butable to its use.   Of domestic ironwork of the time of
Henry VIII., the Tudor fireplaces in the kitchen of Hampton
Court still remain.   In one of these are two sloping brick
buttresses with iron bars of differing length inserted at
right angles and notched to receive the spits.   The same
result is achieved at Plas Mawr, Conway, with two sloping
iron bars and hooks.   There are enormous spit-irons at
Cotehele.   The necessaries for a complete range are enumerated
for the Berwyke Castle kitchen in 1539.   These comprise
" a great chimney of iron, a pair of gallowes of iron for the
same, three crooks of iron, a greate rake of iron, a pan, two
spits, and a small rake."   Trivets are often mentioned in
the inventories of the suppressed monasteries.   Knole
happily retains a great pair of historic fire-dogs, said to have
been saved from the burning of Hever, four feet in height,
ending above in rosettes, five inches in diameter, with
twisted iron edging, the one bearing the Falcon crest of
Anne Boleyn and " H.A." in the characters of the period,
and the other the shield of Henry and " H.R."   These
rosettes are of bronze, all the rest being iron.   Below, on the
moulded shaft, are the figures of Adam and Eve, one on
each dog, skilfully though somewhat rudely forged.   Lower
down is a wide hexagon knop, and beneath this an old stirrup-
shaped drop handle, filled with pierced arabesque ornament
of sheet iron, in which dolphins are prominent.   The feet
are arched, pointed and low, in the Tudor manner, and
strengthened inside by bent trefoils of iron.   These, no doubt,
date from about 1533, the year of Anne Boleyn's marriage.
The attention at that time paid to the designs of andirons
may be gathered from the inventory of Wolsey's furniture
in Hampton Court.   Besides eight pairs of brass with lions,
angels, mermaids, and fools on top, there were in iron six

pairs with Wolsey's arms and hat, four with arms and balls, three with lions, five with dragons, two with balls, one with roses and royal arms. Twenty-two other pairs had the Wolsey arms gilt, with balls; while others had scutcheons, crosses of St. George, double roses, etc. There were fire and fumigation pans, and fire-forks, etc., in great plenty. Henry himself delighted in vanes, and had even his iron bedstead surmounted by four. The roof of his new hall at Hampton Court was provided with twenty-seven vanes by John of Guylders.

CHAP. IV.—THE RENAISSANCE.

ELIZABETH.

FROM the accession of Edward VI. to the death of Mary, religious unrest and attendant events were destructive of art and industrial progress. With the "spacious days" of Elizabeth, who was idolised, a new era seemed to open, persecution ceased, and our own merchant adventurers and explorers were encouraged and foreign mercenaries dismissed. She was not, however, particularly favourable to our native iron industry, viewing with alarm the destruction of timber which this entailed, timber being as essential for shipbuilding and domestic work as iron girders and plates are now. The use of timber was, in fact, prohibited for a time, and she forbade the erection of any new ironworks within twenty-two miles of London, and four miles of Hastings, Rye, Pevensey, or Winchelsea, under a penalty of £10. She, however, incorporated the smiths and blacksmiths into a company in 1578. The German Steelyard, nevertheless, continued to flourish, and Cologne steel was preferred to English, while in 1556 there was actually a movement to close down English ironworks altogether, in favour of Spanish iron, which was then at the extraordinary price of but five marks per ton, the inferior English being nine marks. This must have been merely due to a Steelyard manœuvre, for in 1551 Spanish iron was retailed at £8 10s., and in 1596 at £14 per ton. There were Spanish smiths, too, for in 1556 one was hanged. But finally one of Elizabeth's last and most patriotic acts was to close the German Steelyard for good and all. Flemings had flocked to England in 1576 ; they, however, chiefly occupied

α

Canterbury and Sandwich, where they have left many wall anchors, one of which is dated 1564, and part of another 1603, of patterns such as are seen in Bruges, and formerly at Ypres, in great numbers. There are also many still

FIG. 3.—Wall anchors at Sandwich, Kent.

existing in the old Rows at Yarmouth, some possibly of Elizabeth's date. A very fine lock of this date is on the west door to the choir of St. George's Chapel, Windsor, of three pierced arabesqued plates of sheet iron superimposed and enclosed in cable borders.

In the absence of important outdoor iron gates and railings, the time for which had not yet come, the " grates " to tombs in churches, as they were called, represent the most important efforts of the smith, and in these his progress or the reverse is most readily discerned. But that vast numbers have been destroyed in " restoration," tomb-rails

might have presented an epitome of the average development of this great national industry from the fourteenth
to the seventeenth century. These were not merely protective but in most cases an integral part of the design
of the monument, to which they contribute a certain dignity.
All are formed of bars set vertically and spiked above,
neither confusing the lines of the monument nor impeding
the view, their simple elements acting as a foil and making
the work within appear in all cases richer and more precious.
The vertical bars were for the most part through Elizabeth's
reign still held together above by a cornice, formed as a
hollow box, composed of two wide bars, set on edges back
and front, generally enriched by cable mouldings and
armorial bearings in relief ; and top and bottom horizontals,
to which the verticals below and the spikes above were
respectively riveted. Lofty standards at the angles at first
retaining the moulded buttresses with caps and bases, were
still surmounted by staves twisted, sometimes bearing
banners or merely balls, which add considerable dignity to
the whole. Gilding and colour, often with a touch of heraldry,
supplied a richness, formerly holding its own even amidst
the brilliancy of the coloured windows, but now entirely
lacking.

The earliest existing rail known to me of Elizabeth's
date is that to Lord Willyam's tomb in Thame Church, 1559.
The spikes are moulded in relief, on the face side, into a
well-shaped fleur-de-lis, probably pressed hot into a steel
matrix, and these rise eleven inches above the cornice with
the alternate bars only bluntly pointed. The cornice has
cabled edgings and raised buttons over rosettes, for enrichment, and the standards are buttressed and filed diagonally
to a point without other finial—an unique treatment.

To this or somewhat earlier date may be assigned the
fine rail to Sutton's tomb in the Charterhouse, which retains

D2

the heavy buttressed standards with moulded bases and high twisted staves surmounted by balls, with the cable-edged cornice below, decorated with swans, rosettes, buttons over rosettes, etc. None of this, however, is seemingly at all connected with Sutton. The verticals are javelin-pointed, every third point with the addition of lateral scrolls converting it into a rude fleur-de-lis. The inch bars, set diagonally, leave but one-and-a-quarter inch of free space between them. Nicholas Stone's account for 1615 states that the monument was set up by himself and Mr. Janson of Southwark " for £400 well payed." The rail must be many years earlier, as the present Master indeed suggests, and does not fit its position, nor is it mentioned in the bill, and the Sutton crests resting on the balls finishing the standards are merely clumsy additions in cast lead. The rail to Lady Jane Seymour's tomb in Westminster Abbey, date about 1560, appears in old illustrations with standards without buttresses, finishing above in twisted staves with balls, and with plain spikes over the cornice. Another, to the Countess of Sussex, 1589, was similar, but with fleur-de-lis heads to the twisted standards. In the still existing rail to the Pickering Monument in St. Helen's, Bishopsgate, 1575, the buttressed standards with heavy moulded bases finish in high twisted staves and balls, and the cornice is simply composed of two plain horizontals, the moulded spikes riveted into the uppermost (Fig. 21). That to the Kirwin Monument, 1594, is of twisted bars carried up into balls for the standards, with richly moulded caps and bases: the verticals pass through a single horizontal and end in balls.

A remarkably fine railing screens the Lincoln Chapel in St. George's, Windsor, the burial-place of the Lord High Admiral under Elizabeth, 1584. In this the plain cornice with cabled edges is retained, but the four standards are plain without buttresses, and with heavy moulded bases;

they are surmounted by high twisted staves bearing balls, each with high open-work fleur-de-lis, and a minute acorn for apex.

A richer example is at Ludlow, 1592, with finely twisted verticals carried high above the lock-rails; the standards with fleurs-de-lis and banners at the angles.

The rail to Queen Elizabeth's own tomb in the Abbey, richly painted and gilded, was produced in 1604 by Patrick the blacksmith, to the order of James I., at the cost of £95. The earlier arrangement of buttressed standards and wide cornice was preserved, decorated with " E. R.," but the cresting seems to have been rich with alternating moulded fleurs-de-lis and Tudor roses with their foliage. The *genius loci* perhaps tended to maintain mediæval traditions here which were losing ground elsewhere.

An important departure is seen in the rail to Dean Wotton's monument in Canterbury Cathedral, erected in 1560, six years before his death; the novelty in design due perhaps to the fact that the Dean had travelled much abroad, both as diplomat and counsellor to Henry VIII. and until Elizabeth. In it all traditions are discarded, and the cresting is entirely forged from a single flat bar, placed horizontally, about an inch wide and a quarter thick. This was carried up above every second vertical and hammered into a flat plate on edge some three by four inches over; it was then cut and pierced into an arabesque, and riveted to the short horizontal by half-inch square rivets, the heads beaten flat into pierced hearts. A horizontal, one-and-a-half inch wide and an inch thick, deeply moulded in parallel grooves on both faces, is placed below. The angle standards are merely distinguished by a finial of four acanthus leaves, cut from a stout plate and bent upward to form a nozzle about seven inches high, probably to hold a candle. Another departure occurs in 1593 in the rail to the Jennings monument in Curry Rivell Church in Somerset. Plain and twisted bars alternate, and above is a border of S scrolls, with the central parts hammered flat on edge, and scalloped, fixed to a plate vertically between two horizontals. Flat fleurs-de-lis alternating with flattened spikes form the cresting, and the angle standards are of stouter bars with an open-work obelisk and ball about twenty inches high for finial.

The Elizabethan ironworkers became extremely expert, producing vanes, hanging and wall lights, fire-baskets and grates for burning coal, fire dogs, and even bedsteads. These varied with the localities, the grates, for instance, at Holyrood, Hardwicke and Haddon, Cotehele, Stanbridge, Hatfield and Knole, showing a variety of type and individuality. Ingenuity was exercised in inventing fastenings of all kinds, both to secure rogues — as catchpoles, fetters, handcuffs and collars—and to safeguard houses from depredations with locks, bolts, bars, and chains, some of considerable artistic merit. Also we find, in spite of protests and prohibitions as to the destruction of woods, that Elizabeth herself was a great patron of the ironfounders of the Weald, having commanded at various times a number of models for her royal fire-backs with the royal arms differently displayed, with and without supporters, badges, cyphers and mottoes. Whether encouraged or not, many of the higher nobility failed not to follow her example.

JAMES I.

THE rails to monuments under James I. differ but little from those of the previous reign, the mediæval characteristics being to a considerable extent still retained and even exaggerated. This is especially noticeable in those to which the standards are surmounted by heraldic banners. The most

notable of these is the rail to the tomb of Montague, Bishop
of Winchester, in Bath Abbey, who died in 1618; according to Camden, it cost £300 with the grate. The vertical bars pass through the horizontal, and the spikes are
moulded ; while the effect of the mediæval cornice is produced
by a plate fixed on edge, also moulded with an escalloped
ornament. The standards are of bars two inches square,
with moulded caps and bases, surmounted by a staff, also
moulded, twisted, and finishing in a ball and open fleur-de-lis,
surmounted by a banner with the bishop's arms under
a cross. These finials rise four feet above the cornice,
together with central standards bearing a garter and shield
of arms (Fig. 22). The Hoby rail at Bisham is similar in
general effect, but the bars pass through a plain horizontal,
simply drifted for their passage, without other ornament ;
while the standards are moulded in front only with a buttressed effect, a staff rising above being of baluster
form surmounted by a ball and banner (Fig. 22).

Examples, some of which may perhaps date back to
Elizabeth's time, are at Hunsdon, Herts, about 1618; at
Hawstead, Suffolk ; two at Ashford, Kent ; Whaplode, Lincolnshire ; and Hitcham, Bucks, 1624 ; Chew Magna in
Somerset; and Betchworth, of which latter but one
standard has been saved. The Tanfield monument in
Burford Church, about 1625, has the central point of the
four-way fleur-de-lis of open-work, while on the stem below
are four scrolls with drooping acanthus leaves, the standards
being massively twisted. The monument in Paston Church
to Katherine, wife of Sir Edmond Paston, executed by Nicholas
Stone for £340, is similarly railed, with twisted standards
and fleur-de-lis heads, dating from about 1629. The well-
known grilles of James the First's time, under the arched
openings of the cloisters of Bristol Cathedral, have the
typical two-way fleurs-de-lis on twisted bars, the twist

reversed three times with good effect. The remarkably fine
series of rails in Rochester Cathedral were all made either
in this reign or later. That removed from Bishop de Sheppey's
Chantry is the most striking, on account of the relatively
large size of its somewhat archaistic four-way fleurs-de-lis,
two feet high, and seventeen-and-a-half-inches across, sur-
mounting the heavy twisted standards.

All the rails in Rochester Cathedral have been shifted
about, and thus their dates and original positions cannot
now be known. Certainly none can be of sixteenth century
date, and they may, since they seem to have been produced
at one time, even be as late as about the middle of the
seventeenth century. A rail at Tamworth is also remarkably
fine, and probably of the time of James I. The standards
are twisted, with a cluster of four scrolls and spikes, scrolls
and egg-shaped finial ; the horizontals are plain, and the
verticals finish in javelin spikes ; a central bar on either
side bears four twists welded into a spike. Sir Edward
Hext's tomb, 1623, in Low Ham Church, Somerset, has a
rail with plain and fleur-de-lis points alternating. The gates
at Cowdray, hung from the buttressed standards and perhaps formed from some dismantled sixteenth-century tomb-
rail, may be of old material worked up. A pair of iron
gates closing the fore-courts of Cote Bampton, Oxford, were
no doubt made in King James's reign. The introduction of
fore-courts necessitated gates, but these were for a time entirely of wood, and later of wood frames with panels of
vertical iron bars above and below the lock-rail. One such
is at Groombridge, of the date of the building, 1625. The
panels contain plain vertical bars, which fall toward the
centre, with wood horizontals and a cresting of iron spikes
and fleurs-de-lis. A second pair of gates of the same outline
are wholly of iron.

The outer gates to Blickling Hall, Norfolk, 1626, are of

timber, but their semicircular head is framed to receive
iron fleurs-de-lis.

Ham House was built by Sir Thomas Vavasour in 1610,
for the Duke of Lauderdale, like a great prince's villa in
Italy, with courts, statues, avenues, fountains and parterres.
The iron gates to the fore-court are probably contemporary,
and appear unique. They are lofty and impressive, consisting of four equal panels, the inner opening as gates, and
the outer fixed ; each is of thirteen vertical bars in massive
frames, passing through two horizontals dividing them
into nearly equal panels. A peculiarity of construction is
that the verticals are not tenoned, but flattened at each
end and riveted to a rebate in the frame. There are no lock-
rails, the fastenings being invisible, and no dog-bars, but
a row of plain spikes form a cresting above.

CHARLES I.

THE sons of James were extensive patrons of the arts, and
Dutch and Flemish influence was strong during the reign
of his successor. French influence was mainly represented
by the Gascon Solomon de Caux, who had been their drawing master, and Briot, who engraved the State seals. Bernini
executed bronzes. Among British artist - craftsmen were
Inigo Jones and Nicholas Stone. Our real renaissance, the
near approach to modern conditions, is due to Charles I.

Of important smithwork of this reign there are but few
remaining examples, yet gardens and their planning were
already receiving much attention. That to St. James's Palace
with the Mall, laid out by the Sieur de la Serre in 1639,
had a great number of stone and bronze statues, nothing
being too expensive for the King, and they were brought
from all parts, as to a fair where there is always a successful
sale. In the gardens of Worcester Park, according to the
Survey of 1649, a stone balustrade led up to iron gates,

and there were several others between piers. No garden
gates of this date, however, are known to exist, but there
are gates to Hereford Cathedral, which seem to be of this
reign, on much the same lines as those to Cirencester Church,
but with the addition of a lock-rail of scrolled rosettes
between two bars, and a scrolled cresting to match. A pair
at Llandaff is similarly treated, but the cresting is replaced
by a rosette and scroll to the back stiles. Seventeenth century drawings, as of Sarsden, built in 1641, show gates of
considerable variety, but of timber framing with panels of
plain and twisted iron bars, mostly with some sort of fleur-
de-lis cresting. An interesting balustrade to the buttery
steps of Wadham College of this date has been illustrated
by Mr. Aymer Vallance,* consisting of vertical bars partly
twisted, with scrolls in pairs between curious flower-like
cradles, and fleur-de-lis finials over some of the verticals.

Balconies of iron seem to have been introduced into
England by Inigo Jones in the building of Kirby Hall,
Northants, in 1636. The stone-work is supported on massive
scrolled corbels built in, and the iron rail, undoubtedly
contemporary—the idea probably reaching us from Italy—
is perfectly plain with balls over the angle standards. Other
examples to houses by Inigo Jones are at Thorpe Hall,
Peterborough, and Hutton-in-the-Forest. There are two
balconies at Bolsover of plain and twisted bars, with a pair
of twisted standards at each angle surmounted by balls ;
and a third semicircular in the ruined part, which can
hardly be of much later date. Lord Arundel is credited
with the introduction of balconies into London in 1650,
but their use did not become general until shortly before
the Restoration. The Sieur de la Serre, employed by
Charles I. to lay out his gardens, speaks of " shops, balconies
and windows covered with tapestry," on the entry of Marie

* Aymer Vallance—*The Old Colleges of Oxford.*

de Médicis into London in 1638 ; and the flat roofs of the houses on London Bridge are clearly seen in old views to have been railed with iron previous to the Great Fire.

Of tomb - rails, the most modern looking is that of William Wentworth, Earl of Strafford, 1645, in Peterborough Cathedral. The bars are scrolled at the base and arched to meet above in pairs, with equal spaces between, coupled by moulded collars, with small acanthus leaves beneath. Above is a cresting of scrolls, also with acanthus leaves, to which an earl's coronet forms the central feature. Other tomb - rails, however, differ in no important respect from those of the previous reign. Thus, the tomb of Sir Richard Scott in Ecclesfield Church, 1638, is of alternately plain and twisted bars, stopping beneath a boxed - out cornice, as in mediæval constructions. The twisted bars bear an open spike and light scrolls, suggesting a fleur-de-lis, and the plain bars end above in twisted spikes.

One of the most sumptuous monuments of the reign incidentally introduces an entirely new development of ironwork. It is that of marble and bronze erected by Princess Frances to the Duke of Richmond, in 1639, in Henry VII.'s Chapel at Westminster. The pierced canopy is of rich arabesque design in moulded frames, four-sided, and domed to support a winged figure of Fame. The four curving panels are of stout sheet iron, riveted together, pierced and richly gilt, introducing different mottoes under coronets, crests, and monograms. This treatment may have been suggested by the contemporary Flemish work, frequently welded up of flat iron, like mediæval scrolled hinge-work, by this time almost disused except for pierced casement window-fasteners, work at which the Flemings were adepts. Heavier work of the kind is seen in the wall anchors at Yarmouth or Sandwich, the earlier of which are no doubt actually Flemish productions.

Smiths so expert, versatile, and industrious as the Flemings naturally found work, and it seems almost certain, from the " wall anchor " design, that the pair of iron doors to the side chapel at Farleigh Castle, made for Lady Hungerford in 1650, are by Flemish smiths. They consist of four panels exactly alike, in which Flemish design is distinctly present, with narrower panels holding shields of arms in place of lock-rails. They are disfigured by cresting composed of Hungerford crests repeated in a row and out of scale.

Hour-glass stands were seemingly introduced by Archbishop Parker in or about the year 1559. An existing specimen bears an Elizabethan date, and another of her reign is represented on the frontispiece to John Day's Bible, 1569. Many others of about that date exist, and also of the reign of her successor, but exact dates of works by village blacksmiths are difficult to determine. Dr. Cox's earliest reference * is 1598, to Ludlow, for the making of a frame for the " houre glasses." At Leigh, near Penshurst, there is an actual cradle with the date 1597. The stands usually consist of a cradle somewhat in the form of a cresset or fire-basket, either fixed to a bracket or more rarely to a standard on the pulpit, as in St. Michael's Church at St. Albans, or Flixton Church in Suffolk. In the former the standard is a single vertical bar twisted, to which four scroll brackets are attached supporting a cradle with small C-scrolls pinned to it for decoration. The usual form of cradle is of two rings connected by vertical bars, either simple or decorated. Many are bracketed out from a pier, wall, or pulpit, and either fixed to a plain horizontal bar, generally supported by a stretcher simple or scrolled, and made a vehicle for some simple or rather rich decoration of scrolls, twists, or foliage ; or the bracket is sometimes bent sharply upward at

* J. Charles Cox—*Pulpits, Lecterns, and Organs.*

right angles, giving a vertical support usually twisted, and
frequently with additional scroll decoration. The verticals
of the cradle, seldom more than four, are either welded to
the stem, when one ring above completes the cradle, or
may be riveted to cross-pieces or a plate, when a second
ring is added. In this case the independent verticals may
be balustered or beaten into escalloped or vandyked plates
or discs, and end above in fleurs-de-lis or quainter shapes,
some one hundred and fifty being known. No two are
precisely alike, some are low and almost bowl-shaped, while
others are tall and lantern-shaped. One of the most remarkable is at Stoke d'Abernon, with hexagonal cradle, panelled
with sheet iron pierced geometrically, with scrolled ornament below, and a scrolled bracket.

Hour-glass stands fixed on stems for pulpits are also in
a few cases elaborately scrolled. A fine specimen at Sel-
worthy, Exmoor, has four supports welded to a twisted
stem carried high above the ring; and in another, at
Walpole St. Andrews in Norfolk, the stem opens as a
sheath. To this group belong the hour-glass stands of
Hurst and Binfield Churches in Berks. The Hurst bracket
is a branch of oak and holly with leaves and acorns or berries,
and the letters " E. A.," with date 1636, together with a
minute lion and unicorn. At Binfield is a branch from which
other branchlets diverge in three directions, with twigs
bearing varied leaves, such as oak, trefoil and thistle, with
tulip flowers, and minute lion, swan and bear. The late Mr.
Vincent Robinson had almost a replica, but with vine-leaves
and fruit. Hinges at Stokenham, with the date 1636, and
others, are also of the Flemish school. A pulpit lectern in
Clyffe Pypard Church, of 1629, is interesting, the desk
consisting of a pierced plate with circle and star as the centre
of a lattice pattern bordered by scrolls. We may possibly
owe the richly worked covered lantern-shaped hour-glass

stands of Compton Basset Church, Wilts, and St. John's Church, Bristol, to the same source, as they are constructed almost entirely of beaten plate-iron. The iron holders are of balusters in silhouette; that of Compton Basset is domed and formed of arabesques, ending above in large four-way fleurs-de-lis; and the Bristol one finishes similarly in four buttress scrolls on edge, and meeting at the apex in a flower. Both are, or were, carried on simple iron brackets, comprising a large fleur-de-lis. Of this age is the cradle for an hour-glass at Bloxworth, Dorset, with two tiers of fleurs-de-lis between wide bands; and there are richer examples at Cowden, Kent, and Chilton, Bucks, figured by Dr. Cox. One at South Ockendon, Essex, dates from about 1640. The only one bearing an actual date of Charles I. is here illustrated (Fig. 23). Examplés, possibly also of this date, florally decorated, are at East Langdon, Kent, and Salhouse in Norfolk. In the former the bracket is a twisted bar bent out from the wall and rising vertically to support a plain cradle with a richly decorated stay-bar, forming, half-way up, a cluster of flowers and foliage. In the latter the bracket, on assuming the vertical, breaks into a large lily, from which issue other lilies and three intertwining twisted bars, welded to form the plainer cradle. Another Norfolk example represents a stem with root and foliage passing above into a foliated cradle. Some of the Scottish hour-glass stands are also remarkably interesting, particularly those in Linlithgow and St. Cuthbert's Church, Edinburgh. Similar brackets, but designed to support baptismal basins, occur in several Scottish churches, and good specimens are in the Edinburgh museums.

It is not so very many years ago that two lists were published, each of about a dozen examples, containing all that were then known. The latest list by Dr. Cox contains one hundred and three, but omits eleven of those in the older lists, which have presumably disappeared. I am able to add fourteen, bringing the total, exclusive of Scottish examples, to about one hundred and twenty-eight, less any that may have been lost in the interval. These are of iron, but one cradle is on a carved wood bracket, at Cliff in Kent, dated 1636, and another at Waltham, Leicestershire, is on a wood standard.

Every church standing during the Reformation to the time of Charles II. must have possessed one, and no doubt more will be discovered. Some were of brass, like that still existing in St. Alban's Church, Wood Street, or the fine example in the Danish Church in Wellclose Square, with four hour-glasses. One, of massive silver, belonging to St. Dunstan's in Fleet Street, was melted to make the beadles' staves.

Font cranes, a purely Flemish idea, are represented in St. Alphege at Canterbury, worked in the Flemish way, of iron scrolled on edge ; and, with the more important one in the Cathedral, between two brackets and dated 1636, are clearly relics of the Flemish influx to Canterbury, described by Sully in the time of James I., to avoid the terrible religious persecutions following the conquest of Belgium by Spain. The unique iron cradle in the Ashmolean Museum at Oxford is of the same work. No doubt equally Flemish in conception and craft is the well-known Dartmouth Church door, overlaid with iron, beaten out into leopards, *passant regardant*, part of the ancient arms of the town, backed by a tree with root and branches symmetrically disposed, perhaps representing the vine as an emblem of the Church. Iron figures give the date 1631, and each vine-leaf is nearly a foot long. Iron was in common use throughout the reign for all general purposes, such as stair-rails, brackets, scrapers, chests, and particularly vanes, some of which are very fine. Little was produced under Cromwell, when art work was practically dormant for ten or twelve years. Five years were absorbed by Civil and the Dutch wars, and his autocratic power as Protector barely lasted so long. Sturdy simplicity was practised in the troubled times, but gave way to richer fashions in the last few years of power. Balconies began to become popular in London while Richard was Protector. The first actually recorded was at the corner of Chandos Street, " which country people were wont much to gaze on." By 1659 " every house " in Covent Garden " had one." The possession of a balcony, or its colour and gilding, became distinguishing signs to houses, when none were numbered, and citizens were hard put to it how to direct strangers to their abodes.

CHAP. V.—RENAISSANCE DOOR FURNITURE.

HENRY VIII. and Louis XIV. were both supremely kingly men, and each at the outset of his career had the inestimable advantage to be tutored by an astute and gifted counsellor, the one by Cardinal Wolsey and the other by Cardinal Mazarin, through whom their respective countries were in turn raised to great power and prosperity. In their reigns feudalism in each case was finally swept away, civilisation distinctly progressed, and the personalities of the two monarchs thus became indelibly impressed on the peoples they ruled.

Under Henry VIII. fortified and monastic dwellings in England gave place to spacious and sumptuously decorated houses and palaces, environed by gardens and pleasure grounds. The house, hitherto a place of safety or shelter, definitely became a home suited to the status of the possessor with all necessary amenities and comfort.

Louis XIV. had chiefly political tasks to contend with; but, living in more advanced times, he was able to introduce into France still more modern and palatial dwellings, with spacious staircases of stone and richly wrought iron balustrades, as well as grandiose iron entrance-gates to palaces and parks; whence the fashion quickly passed into England.

Here the changes inaugurated by Henry were developed under Elizabeth and her successors. Mansions appeared with stately suites of reception rooms, abundant sleeping apartments, studies and libraries, all with ample windows; comfortable beds, tables and chairs, replaced trestles and benches. These changes permanently established, attention

E

could be turned to minor details, in regard both to comfort and decoration. Under James I. iron grates for
burning fuel, no longer exclusively wood, were installed on
the spacious hearths, with iron or gilt and enamelled brass,
or even silver fire-dogs; while candelabra of iron, wood,
brass, or silver depended from the ceilings. But in nothing
is the excessive attention to detail more conspicuous than
in the window catches or fastenings. These, of pierced and
hammered iron, are in endless variety, single, or sometimes
duplicated for larger casements, and connected by a finely
moulded attachment bar worked by lever handle. The
large number at Haddon Hall, accessible to the public,
are probably the best known, and in a general way the most
typical. Hardwick, a somewhat earlier building, is less
rich, but its window-catches are of interest, as pierced with
the intertwined serpents of the Cavendishes. Elsewhere
a few take the form of animals, such as the greyhound at
Banbury Vicarage; or of fish, as the hippocampus at Cars-
well House, near Faringdon, and the dolphin in Charlecote
Gate House. The vast majority, however, are cut and pierced
in the arabesque of the period, and this in infinite variety,
for of many score not two are alike. The mechanism is
usually a lifting latch held by a spring, actuated by a handle
or ring, and necessitating a fairly long plate, broad at the
extremity and tapering towards the fulcrum. The latch is
shaped into some semblance of a fleur-de-lis, fantastic
partisan or lance-head; the plate being cut and pierced
into dolphin or such-like silhouettes.

Guildford Hospital preserves many of these old window
and door fittings, and, in fact, no Jacobean house, unless
completely modernised, is without them. In a slightly
later type without spring, the catch is lifted by a long and
slender rat's-tail handle, with a scrolled end or knob carefully weighted to balance. Smaller casements are opened

by a twin buckle or ring on a pierced plate, and in some cases a bunch of light-welded scrolls takes this place, while a sliding bolt may do duty for the lifting latch. The back ends of the plates are sometimes continued as rods across the casement to form a stay, and the fixed parts of the gear are welded either to the casement or its frame. Thus it is apparent that the actual smith carried out both utility and purely decorative details. Lifting latches for doors are stronger and usually more simple than those for casements, but in some few cases their fixing-plates are also richly pierced. The great lifting bars and fastenings of the portals of colleges in Oxford and Cambridge, and of a few churches and mansions, are also interesting relics of bygone ages.

Door-hinges of Jacobean interiors are of the " H " type, either of twin vertical plates with the ends silhouetted in the outline of eastern domes, or with the same reversed and pierced in the horizontal ends. Generally, however, hinges are of the curving twin " cock's head " design, but even in such the variations appear endless. The perfectly developed " cock's head " somewhat faithfully represents that of the crowing cock, all the four terminations of the hinge being similar ; but this bird's head becomes so variously interpreted that the beak may be closed, the comb much exaggerated or omitted, and the head, in fact, altered to the semblance of either a worm, serpent, or griffin, while the curving breast and neck of the cock may be full or meagre, sharply or slightly bent, or even angulated. Bodies of hinges are either cut back to a cusp or trefoil, or cut into a projecting halberd, javelin point, or fleur-de-lis. The Bodleian at Oxford possesses some in which busts of elderly men, apparently lecturing or preaching, form the terminals, and others at Haddon are reminiscent of animals. Small strap-hinges, for aumbries or cupboards of this date, are

E2

usually narrow, angular in outline, and of substance permitting parts to be beaten out flat to form projections and terminals of trefoil or more elaborate design. Made smaller, similar designs serve for casement-shutters with double joints, or for aumbry doors. External door-hinges are much stouter, with javelin points, sometimes with lateral embellishment such as scrolls, tridents, trefoils, or fleurs-de-lis. Hinges of diverging and recurved scrolls afforded secure fixing, and might terminate simply, in leaf-points or animals' heads. Examples are at Haddon. At times a badge, as the Pelham buckle, may be included in the hinge design.

Bolts, whether long and heavy for doors, or short for cupboards, offered opportunities for design, as at Hardwick and Haddon. An unusual number, with other door furniture, is in the Museum at St. Albans. The "rat-tail" bolt, enabling the bolt to be reached, is also a favourite; a particularly fine example is on the inside of the entrance door to the Jerusalem Chamber, Westminster Abbey.

Devices to secure and safeguard alms-boxes in churches deserve notice, the more so if with original padlocks, which take varied forms. Also hinges and locks of muniment chests, and those to vestries and private chapels, stairs, and other old doors, whether in churches, cathedrals, colleges, or mansions, are often of great interest, not only to the mechanic, but to the antiquary and artist. Locks are sometimes dated, those in the hospital at Guildford having 1619 or 1620 cut on their bolts.

Some comprise a rebus, badge, or initials: they repay research; but all seem to pale before the great gilded iron lock of Henry VIII., from Beddington, a personal relic, with the royal arms chiselled in relief, with vertical bands of varied tracery. This specimen, now in the Victoria and Albert Museum, is unique, a great national and historic treasure. It must be confessed, however, that even this royal lock, made for an all-powerful and extravagant monarch, who never hesitated to employ skilled foreign craftsmen of all nationalities, does not equal in " *ciselure* " the almost goldsmith-finish of the finest Parisian craftsmen working for Henri II. Though, perhaps, few large English decorative chests of this reign now exist, groups of them, mostly with domed covers, are represented in the Wolsey tapestries of Hampton Court. These probably represent " Standard " chests, as used by Henry in his royal progress. It was enacted in his time that all parish registers were to be kept in coffers with two locks and keys ; and also that jewels and valuables were to be kept in iron chests. The church chests are usually oblong, with raised or domed lids, closely bound with iron straps, having several handles, and either three locks or three padlocks, with rings and staples through which an iron bar is passed and secured. Among historical chests preserved is that in which Domesday Book was formerly kept, banded all over with iron strap-work, studded with nails, and secured by three locks. The lid is deep and oversails the box, like many of those in churches, and is difficult to force. A similar box, with two locks and three padlocks, in which the Scottish Regalia was discovered, is in Edinburgh Castle. A domed example, almost concealed by iron bands, with lock and staples for bars, is in St. John's College, Cambridge, and so heavy that it is mounted on iron wheels. Another, of similar dimensions and form, and also on wheels, is fashioned entirely of thick plate-iron, closely banded vertically with iron reinforcing straps, and provided with four long hinges bifurcating into cusped and trefoil ends. It is in St. Columba's College, Dublin, and of native work. From this it was but a step to plain solid plate-iron rectangular chests. One such, of extraordinary dimensions and weight, is possessed by Colonel Wellesley, of West Green

House, at Winchfield. It seems to have occurred to some
astute eighteenth-century German trader to reproduce chests
of the Domesday type wholly of iron, and to make them an
article of commerce. They are constructed of plates not
very stout, but securely dovetailed and banded together ;
the lock, occupying the interior of the lid, masked by a
pierced sheet-plate of German design, shoots from six to
twenty-four bolts, catching under the iron flange, and thus
securely holding down the lid. Dummy escutcheons are
ostentatiously provided on the front of the chest, while the
real key-hole is ingeniously concealed in the lid and actuated
by a secret spring. Even our Bank of England, City Companies, merchants, and others were customers for these, and
an immense trade was carried on to the detriment of native
industry.

A remarkable church chest is that of St. Giles-in-the-
Fields, in which large escutcheons of the royal arms and
those of certain nobles, of pierced iron, form the terminations
of vertical and wide bands of arabesque design below
the lid, concealing most of the woodwork. The date of
this is 1650.

Escutcheons or guides for the keys were also vehicles for
ornament, being larger or smaller decorative plates nailed
to the chest or door. In size they vary from about six inches
to an inch, and where possible were designed in Jacobean
days to comprise arms or badge of the owner, like the wheatsheaf and sickles of Farleigh Castle, or the Ferrers' arms of
Baddesley Clinton. At Wilne is one surmounted by the
imperial crown, and another by two crowned lions rampant.
At Compton Wynyates is one formed of two dolphins
regardant, and another of two ducks' heads averted. These
latter *motifs* became little less popular in degree than the
" cock's head " for hinges, though so rendered as to
be in many cases scarcely recognisable. Many follow these

types, the fixing nails providing eyes. Few show wide
departures unless specially narrowed to suit particular conditions, or intentionally plainer. Another form is the heater
or shield shape, sometimes with small fleurs-de-lis pierced in
the angles; other designs comprise the horns of crescents.
Very rarely they take the form of an initial letter, as
the " T " of Wickham Court.

The old protective gratings to the book-shelves in the
Bodleian Library are decorative and of finely pierced iron,
now probably unique. Jacobean iron handles mostly take the
form of loops, ellipses, rings or stirrups, often comparatively
slender. These are at times bent abruptly upward or
doubled over, in the lower part, and then hammered into
a trefoil or other simple form, as well as either twisted or
flattened. Thicker handles of square or rounded iron are
embellished with mouldings and knops. All hang perpen-
dicularly from the spindle. Closing rings depending from
two moulded straps are less common and probably of earlier
date, and may possibly have suggested the Scotch tirling-
pin, formed of a staple of twisted iron through a loose ring
fixed to the entrance door or jamb. A brisk friction of the
ring up and down the iron twists produces a strident and
penetrating summons. Many still exist in Scotland, though
the original idea seems Flemish, since more ancient examples
are in Bruges. The Bible chains with swivels in old public
libraries and churches are also interesting survivals.

Of iron knockers the most favoured and oldest design
seems the knightly prick-spur, or still more the rowel-spur,
which no doubt carries back to mediæval times, since they
are found in such venerable places as the Windsor Cloisters,
the Jerusalem Chamber, Nottingham Castle, Brazenose
College, Penshurst, Lacock Abbey, Hardwick, the Bishop's
Palace and the Deanery at Wells, and so on. In these the
striker takes the form either of rowel or pricks. A simple

form of knocker is the hammer, in which the lower end is
sharply bent at right angles for percussion, the shaft being
curved somewhat outward for convenience of purchase
and force. A fine and early example is in the cloister at
Eton, but they are not rare, and some few may be of considerable antiquity. The shaft end is often split to diverge
and be held in a pair of fixed sockets or rings. The shaft of
one of these large hammer-shaped knockers, on the west
door of Canterbury Cathedral, is embellished with raised
button-like pellets, the lowest stamped with a cross. The
shaft of another, on the door of the Deanery at Wells, has
a raised rectangular zig-zag decoration. Other knockers
are fashioned as scrolls with buttons for percussion, while
some are merely rings, or of the various forms taken by
pendent handles, but of heavier make.

CHAP. VI.—CHARLES II. PERIOD.

A DISTINGUISHING feature of the Restoration is the increasing popularity in London of iron balconies. In 1661 the genial Pepys speaks of " balcone windows " to Lord Sandwich's house, and in 1662 he witnessed a show from one " over against the Exchange."

In the Rules for rebuilding after the Fire it was enacted " that All houses to be erected in the high and principal streets shall have Balconies four feet broad, with Rails and Bars of Iron, equally distant from the ground ; every of which Balconies shall contain, in Length, two parts of the Front of the House on which it shall be placed, in three Parts divided ; and the remaining vacancy of the Front shall be supplied with a Pent-House, of the Breadth of the Balcony, to be covered with Lead, Slate, or Tile, and to be ceiled with Plaistering underneath." Iron rails as balustrades might also be put round roofs, permission of the Committee for Rebuilding being obtained.* An early drawing of a balcony to a house in the Strand shows it as prescribed, of plain bars, but with four stout twisted standards and balls over them. The balcony at Bow Church, and others put up by Wren shortly after, are of plain vertical bars. But, with smithing active, decorative elements could hardly remain excluded ; thus, that to the second Royal Exchange introduces narrow panels of scrolls at the angles. One, to a house in St. Albans, is of plain and twisted bars, but with four panels, each of six C-scrolls back to back, attached to the upright bars by riveted spikes beneath a handrail with escalloped edge, and large balls over the

* Stow, Book I. Cap. XXVIII.

standards. A balcony to the Cupola House at Bury St.
Edmunds has similar verticals with three twisted standards,
each with a pair of C-scrolls and large ball finials. Another,
at the Town Hall, Wallingford, said to date from 1677, is
of verticals also with balls over the standards and two
panels of simple scrolls. The best of the older balconies
is in the High Street at Guildford, of alternatively plain
and twisted bars, with cressets over the heavy angle
standards, and a central panel of two horizontal bars terminating in large thistle-heads, crossed diagonally by bars
ending in scrolls and spirals, with a small bird-like head
between tulip-leaves. In 1683 the great clock was presented
to the Town Hall, with stay-bars similarly decorated, but
with thistle terminals. This design is again repeated in the
panels of the Town Hall balcony, erected in 1697.

Few of the seventeenth century tomb-rails have been
spared by church restorers, but several are seen in Drake's
Eboracum. That to Archbishop Frewen's tomb, 1664, was of
round bars, the standards twisted and javelin-pointed.
Archbishop Sterne's monument, 1683, had a rail of square
javelin-pointed bars, but with a detached point between,
supported by scrolls ; Archbishop Sharpe's rail, 1713, was
similar. Of two yet remaining, one is that of Archbishop
Lamplough, who died in 1691, perhaps erected before his
death. In this the stout twisted standards to the rail have
high moulded points and are buttressed by scrolls ; the
bars, similarly pointed, are also with supporting scrolls ;
the horizontals moulded on edge, and the vertical bars
below with scrolls attached at intervals. The other still
extant is Archbishop Dolben's, 1686, of plain verticals
connected by two moulded horizontals and a bottom bar,
and five massively twisted and javelin-pointed standards on
great ball bases ; the rails support a scrolled cresting with
spirals and a shield, and rosettes are attached to the

standards. Elsewhere existing tomb-rails of this period are more severe, one in Romsey Abbey, 1658, being of plain and twisted bars with short spikes passing through the horizontals ; while another, in Ufford Church, near Woodbridge, 1671, is similar, but rather more decorative. A curious reversion to mediæval precedent is presented by the buttressed supports to the bell windlass in the Curfew tower at Windsor, inscribed by " John Davis the smith," 1689.

Other examples of fleur-de-lis headed rails of this date are at Blackwater Church and Canons Ashby Church.

Ordinary outdoor railings of this date are mostly severe, and among the best known and most remarkable are those to the Sheldonian Theatre, and to the Quad of Christchurch College at Oxford. The Sheldonian is in short lengths on a dwarf wall, between stone piers capped by busts, which compel attention. It is remarkable for the height of the spikes, half the total height of the rail, with a central standard carried higher bearing two C-scrolls near its apex. These were noticed by Miss Fiennes, and again in 1688. The Christchurch railings, by Wren, must be nearly contemporary, and are also on a dwarf wall, with unusually long spikes, similarly divided into short lengths by massive stone piers. The centre of each is very happily marked by a beautiful scrolled design carried up above the top horizontal to form a panel between two extra long spikes. The front of Wadham was formerly screened from the street by sixteen stone piers with similar railings and gates between, and there was a somewhat similar rail to Lincoln College. The most interesting example left in London, now but a fragment, closed the forecourt of Lindsay House, built by Inigo Jones, on the west side of Lincoln's Inn Fields. A rail to a tomb of 1692, in Hitcham Churchyard, has plainer moulded spikes with twisted standards. Some of the older railings of Chelsea Hospital may date from

1682, and are worth noting for the moulding of the spikes and the heavy twisted standards.

The tulip as a decorative motive in ironwork is easy to forge, and generally pervades the smith's productions during the third quarter of the seventeenth century. There had been a tulip mania, and the flowers were favoured by painters, embroiderers, potters, and silversmiths. Also artists of repute from the Low Countries swarmed at Court, the number having risen from only five in the time of James to twenty in that of Charles I., and to thirty under his successor. Tulips shared the popularity of the oak, thistle, and rose after the Restoration. In the hands of the smith, however, it sometimes reverts to the better-known iris or fleur-de-lis, no longer popular as an emblem. An instance of this is seen in the remarkably interesting hour-glass stand of Easthope Church, near Much Wenlock, of scrolls and silhouetted balusters, with a small banner in front, indented and fretted with arabesqued fleurs-de-lis and the date 1662, also pierced with " S.S." and a heart, the staff surmounted by an open iris-flower. Perhaps the latest of the standards to the railings in Rochester Cathedral, surmounted by two clusters of four scrolls on a spike ending above in a tulip, may be of this same year, when Merton College was restoring the tomb of Walter de Merton. A mural monument to Sir Robert Lloyd, 1676, in Wrexham Church, preserves the remains of an overthrow of iron with tulip-flowers and foliage. There are two hat-stands in St. James's, Garlickhithe, rebuilt soon after the Fire, each comprising two tulips with an oak-leaf and acorn, amidst barbed arrow - points, the connecting scrollwork hollowed into a half-round section. At Penshurst a wide pair of low spiked iron gates have tulips welded to the back standards. One of the entrances to Ham House has an over-throw of scrolls supporting a bunch of five tulips over the gates. These, like the older pair, are of four panels, of equal width, two opening as gates ; all are of vertical bars with. fringes of S-scrolls above and of C-scrolls with waved points both above and below the single lock-rail, with crestings of plain spikes, together with arrow-pointed dog-bars— a very early example of these—supported by scrolls at the base. The third pair of gates at Ham House, closing the garden approach, are relatively narrower, occupying but a third of the total width between the stone piers. Except as to the double horizontals, they are identical with the rest, apart from the cresting, which is of clusters of barbed arrow-points, interrupted over the gates by a scrolled overthrow supporting a roundel of arms in a garter, under a ducal coronet. Such entrance-gates to the spacious approaches or forecourts were favourably placed for displays of heraldry, notifying the *status* of the proprietor. Heraldic bearings, monograms, and coronets, soon became the fashion over gates to mansions and country seats. To revert, however, to tulipwork, the most striking of existing examples is near the Castle at Windsor, where a grille under an arch comprises eleven large interlacing scrolls clothed with tulip-leaves and flowers, around a monogram " G.A.R.," possibly of George of Denmark and Anne. It rests on a border of Greek wave-scrolls, and is surmounted by a small but carefully worked finial with scroll supports. No gates exist, but some dealer contributed to an exhibition, held more than forty years since, two extremely rich clusters of tulip foliage and flowers springing from many-petalled rosettes, the welding points concealed by formal acanthus, perhaps wreckage from these gates. Other pieces show peculiar spirals and rosettes, which we shall see are frequently associated with the tulip in Scottish work. In England the association is rare, and only known in two remarkable crowns in Leybourne Church, Kent. They are arched over, one with four tulips springing from the diadem, and

the other with fleurs-de-lis and roses of eighteen petals.
Crowns formerly associated with helmets in churches,
as at Penshurst, or with tomb-rails as at Peterborough,
and with sword and mace rests everywhere, are not uncommon. A fine cresting with tulips and spirals to wooden
gates near Greenwich Park was sketched some years since,
and one with tulip-like leaves from Owlpen, Gloucestershire,
has been published. The cresting to the gates of Brewers'
Hall with tulip-like clusters of leafy spikes alternating with
fleurs-de-lis is familiar to Londoners, and there is a somewhat similar cresting to the Newark Church gates. Wren's
fleur-de-lis cresting to the screen doors in Middle Temple
Hall, 1697, is familiar, and there are many other examples
with fleur-de-lis spikes.

Curious gates of Charles II. time are at Queen's College,
Oxford, illustrated by Mr. Aymer Vallance. Each gate, massively framed, is a panel entirely occupied by bars crossing
diagonally to form a lozenge design, supporting a central
oval of four bent scroll-ended bars with long tapering and
twisted spikes at the meeting points, similar spikes breaking
into the lozenge spaces. The narrow pilasters and overthrow
are also treated somewhat geometrically with scrolls and
spikes. The geometric effect is not unpleasing. A far larger
and richer gate of the same college has disappeared. Wadham
possesses a wicket of unusual construction, curved bars
intercrossing to leave a central vesica and group of scrolls
on either side. A pair at Staunton Harold Church, of 1663,
are severely plain, only relieved by the fleur-de-lis ends of
the bars. A singular gate, brought to Hatfield from Sir
Francis Butler's Park, who died in 1691, consists of a wicket
and side panels including small oak-leaves and acorns in
its scrolls, unusual dog-bars, and a straggling overthrow with
monogram and repetitions of the oak. Though skilfully
forged, the design is weird and local. The thistle is seen in

the Guildford balconies, but the richest and best-preserved
example of this, in combination with the rose, is in the panel
of a wooden door in the cloister of All Souls' College, Oxford.
It presents four groups, comprising a central thistle flower and
leaves, with a pair of diverging Tudor roses emerging from
thistle-leaves below, fitted into heart-shaped spaces, two
upright and two reversed, while a lower panel is of plain
and twisted bars. This probably dates from the Restoration. The thistle is not common in English ironwork, but the
rose, no longer Tudor, but of a multitude of petals, is of
frequent occurrence. A fine example is seen in the cresting
to the wooden gates of the Warburton Chapel of St. John's,
Chester. The cresting is of fleurs-de-lis and javelin-points
with lateral scrolls, but the side panels are each of a single
twisted bar, ending in a seeded rose eight inches in diameter
with five concentric rows beyond of petals in outline.
Over the centre of the gates is a group of convoluted scrolls
and spirally twisted tendrils, and the panels below are of
twisted bars with heavy javelin-spikes between. A railing
in the same chapel, of twisted bars and standards with
curious pierced javelin-points and fleurs-de-lis, has a large
nine-inch rose on the central standard. Similar fleurs-de-lis
and spiral twists are preserved in St. George's, Southacre,
Norfolk. Another of roses and spirals, by the same hand,
is in the Dolben tomb-rail in York Minster, dating from 1686.
The Warburton family were having similar work carried out in
their chapel somewhat earlier in the century. Work of this
design was prevalent in the central links to chains in churches
carrying heavy chandeliers. These usually centre in a large
rose attached to the twisted suspending rod, with hollowed
diverging leaves or scrolls in pairs, in which the long spiral
twists are housed, while from between them spring roses,
and sometimes tulips. Three fine specimens of such links
are in our museum (Fig. 24).

Specimens of similar work may be studied in the sword and mace rests of City churches, such as Allhallows, Lombard Street, where there are two ; St. Andrew's by the Wardrobe ; St. Stephen's, Coleman Street ; St. Margaret's, Lothbury ; St. James's, Garlickhithe ; St. Mildred's, Bread Street ; mostly churches by Wren, some built soon after the Fire of London. Brackets to font-cranes of the same work have mostly been disestablished, but there yet remain a few of the rare hat-stands with central rosettes. Hourglass stands were dealt with in Chapter IV. A more remarkable link for a chain, from St. Stephen's, Walbrook, completed by Wren in 1679, is in the Guildhall Museum, with the spirals, oak-leaves and acorns. Wealth, and overwhelming changes of fashion introduced on the accession of William and Mary, must account for the paucity of ironwork of preceding reigns remaining in England.

It was otherwise in Scotland, where the dethronement, banishment, and sudden extinction of the Scottish dynasty must have been deeply resented. Little change in fashion is thus seen in the ironwork across the border, while it underwent a complete change here. In Scotland hardly any mediæval ironwork exists, and the earliest decorative work remaining appears to be the stair-balustrade of two flights and a short landing-piece at Holyrood, erected between 1671 and 1679. The stair panels are hammered, like the older English platework, each forming the royal monogram " C.R.," twice repeated, under a crowned thistle-flower and two leaves. The design did not fit the landing, and the " C.R.' design is supplemented by two narrow panels of older wall - anchor pattern, while the fixing-plate might have been picked up in Bruges or Ypres. It must have been forged by a Flemish craftsman, probably who was responsible for the almost contemporary work at Farleigh. The newel is a plain bar surmounted by a ball, and the iron hand-rail of a small moulded section. Next in date is a rail to the grave of Archbishop Sharpe, in St. Andrews, dramatically murdered in 1679. A real protection was needed, and thus the rail from the ground is over eight feet high. No demand for decorative ironwork in Scotland had hitherto appeared in connection with architecture, and no changes introduced by William penetrated there, and thus, where gates and railings to modern mansions and forecourts were needed, the old Stuart designs were requisitioned. The work at Caroline Park House, Granton, so named after George II.'s queen, is perhaps as early as any. A stair balus-strade of large and boldly convoluted scrolls in the grand manner and once gilded seems to have been by a local smith on the lines of an Italian plaster-ceiling They are forged from a round iron section and clothed with acanthus leaves, scrolled tendrils and rosettes. The work is somewhat rudely executed, and the construction mainly by bolts and nuts. A small balcony to the porch displays the coronet and monogram of Viscount Tarbat and his lady. Panels on either side include family crests and the rose and thistle, emblematic of the Union the viscount had so much at heart. These date from 1696, while a second stair balustrade may date from 1685, in which C-scrolls welded together, and supporting a branch of either rose, oak, or thistle, form the panels. The gates, now removed, probably also date from 1685, and are of plain bars under a frieze, with tulips and medallions of thistles. The scrolled overthrow comprises roses, tulips and spirals, with a central shield of arms and high spike of tulip-leaves. Gates to Traquair House, Peeblesshire, have a similar overthrow with tulips, roses, and spirals, and are of vertical bars crossed by three horizontals with varied fringes. These form part of a fore-court enclosure one hundred feet wide, the railings, on a dwarf wall, having fleur-de-lis heads with waved spikes. Simpler gates to the Avenue have

F

railings at either side, each with a single high central standard
bearing a large bunch of tulips.   A stair balustrade at Craigie
Hall, Midlothian, is like the older one at Caroline Park, but
with realistic sprays of tulip, rose, or thistle depending alternately from the centre of each panel.   The landing panel
of scrolls bears the coronet and monograms of the Earl and
Countess of Annandale.   The hand-rails are in all cases of iron.
A small plain pair of gates at Hopetoun House, Linlithgow, is
notable for the splendidly welded and forged tufts to the
spikes, like iris flowers and leaves, and finely twisted bars.
A short railing at Craigie Hall is no less interesting.   Other
gates deserving attention are at Innes House, Elginshire,
and a stair balustrade in the old Town Hall of Dumfries.
The most stately array of ironwork in Scotland, however,
is that to the terraces and steps at Donibristle, descending
to the Forth, the central feature being a large arch of latticed
and trellised iron with leaf clusters, rising from stone piers,
and surmounted by a high lantern-like structure sheltering
an " M " under a coronet between leonine masks.   Crowning
this is a fine finial of tulips, also repeated over the stone
piers.   Almost all here mentioned have been illustrated by
Mr. Bailey Murphy.*   There is little, if any, further work
of this character to chronicle north of the Tweed.

* Bailey Murphy—*English and Scottish Wrought Ironwork.*

CHAP. VII.—TIJOU AND HIS WORK.

THE year 1688 inaugurates the change of dynasty and much else that matters to our subject; for in this year the fashion of decorative smiths' work commenced to change radically. Some such change must have come about in any case; but that it was immediately due to one man is the important fact, and can never be forgotten. Our ironwork had leaned to the Dutch, for Dutch and Flemish artists were most in favour at the Courts of the Stuarts, and increasingly so during the reign of Charles II., when they permeated society. Charles and James, however—one from the love of luxury, and the other on religious grounds—personally favoured France, and French influence had never been wanting at Court. De Serre and Le Nôtre were employed to lay out the royal gardens; Blondeau and Petitot were smith-masters, who, together with Chéron, Laguerre, Le Sueur, Simon, Roettier, and many others, kept French art influence from fading out. Marot became William's Court architect, and made the designs for laying out the gardens of Hampton Court, which he had decided almost from the first to make his country residence. Tijou appeared on the scene at a happy moment and became introduced to Queen Mary, who took a keen personal interest in laying out her new gardens, and he enjoyed her gracious patronage to the fullest extent while she lived. His fine folio of designs, engraved by the best artists of the day, presents her Majesty on the frontispiece in the guise of Minerva reclining, attended by Vulcan, Mercury, Fame, amorini, and a group of the Arts, in the approved French style of the day. Within a year of William's arrival, Tijou

F2

had set up forges at Hampton Court, and rendered his bill
for six iron vanes " finely wrought in Leaves and Scroll-
worke," £80, and for the rich balcony to the Water Gallery,
the Queen's abode while the Palace was building. This was
taken down to improve the view in 1701, and all record of
these, the first works of Tijou in England, is lost. Miss
Fiennes relates that this water gallery opened into a balcony
to " ye Water and was decked with china and fine pictures
of ye Court Ladyes drawn by Nellor." The second bill,
in 1690, was for the celebrated garden screen with two great
iron gates and wickets, eight square pillars of ornaments,
twelve panels and ten pilasters between them, for £755 7s.,
made for the " Fountain Garden." This costly screen was
hurried on to afford privacy to Queen Mary when laying out
and planting the garden, a large semicircle on the east
front, still known by the name, and overlooked by a park
accessible to the public. Already, in 1689, Evelyn notes
that this was being laid out as a garden with fountains " at
the head of the canal " ; and Gibson's note on gardens, 1691,
describes it as a large plat " environed with an iron palisade
round about next the park, laid out with walks." Miss
Fiennes' diary of the same date is explicit : " The gardens
are designed to be very fine, great fountains and grass plotts,
and gravel walks, and just against the middle of ye House
was a very large fountaine, and beyond it a large canal
guarded by rows of even trees that run a good way. There
was a fine carving in the Iron Gates in the Gardens, and all
sorts of figures and iron spikes round on a breast-high wall
and several rows of trees." Mr. Law, in his work on Hampton Court,* says that the screen and gates were made for the
small " Privy Garden on the river side," but no carriage
entrances could have been of the slightest use on this site,
and he adds they were subsequently removed to the

* Ernest Law—*The new guide to the Royal Palace of Hampton Court.*

Fountain Garden. This view is quite untenable. The screen was made for and erected in the then relatively small Fountain Garden, but was removed five years after the Queen's death, when the garden was re-designed by Daniel Marot, *Architecte de Sa Majesté Britanique*, and enlarged up to the canal head with an outer row of eight more fountains added, entailing a fresh iron railing of no less than 1,442 feet to enclose both sides of the garden. The screen, no longer of use, was then removed. Only the gates were retained and fixed in the same line, but much further from the Palace. The screen itself was also refixed in the old Privy Garden, when £832 was spent in masonry for foundations. A view in Sutton Nicholls' work* makes it clear that no screen was there when the Queen died. In 1783 it reappears as " the magnificent gates and rails of iron, parallel to the Thames for six hundred yards, broken at intervals of fifty yards with twelve gates, 4 yards wide and seven feet high," evidently .the twelve panels of the screen.

A note in Sir John Soane's handwriting records that one set was sent in 1825 to Bruton Street, and the original " Foliage, Ornaments, Bands, etc., of sheet iron, were reinstated with copper, while their appearance being thought too low they were rose one foot." This fact was kindly communicated by Mr. Walter Spiers, who adds that Sir John was unable to learn where they were made, but reported their weight at five tons one hundredweight. These are now between the piers prepared for much larger gates under Queen Anne. The second pair was taken from the riverside and replaced in its second position along with a Birmingham-made copy. The most beautiful panel is fittingly the Royal Rose of England, of many petals, the centre embossed, the rest of open ironwork with the outlines of petals

* Sutton. Nicholls—*Prospects of the most considerable Buildings in and about London. 1724.*

radiating beyond, and with the forged sepals and foliage filling the panels. The Scottish centre panel was a large thistle-flower in a group of ten finely modelled leaves, but is now a flower on a thin curving stem, with travesties of five leaves and tendrils! With this Tijou could hardly have been concerned. The fine female caryatid of the Irish Harp, on the contrary, is perfect, except as to a few leaves of the supporting scrolls. The graceful William and Mary monogram is also probably original; but the Star and Garter centre, the only one not illustrated in Tijou's book, is commonplace, with iron bars radiating from the Garter filling the entire panel. Tijou would, no doubt, at least have preserved the star shape of the rays, and probably have introduced some features of the Collar for the rest. He evidently intended to concentrate patriotically on the fleur-de-lis panel drawn to a large scale, supported by winged caryatids and a glorified pedestal, with garlands, masks, and foliage, one of the most imposing designs in his book; but in execution the fleur-de-lis was mutilated beyond recognition, the King probably not caring to have the arms of his then arch enemy flaunted in front of his palace. Each design occurs twice in the twelve screens, like the fine compositions above them, which are among Tijou's *chefs-d'œuvre*. These centre in masks, five of which are engraved to a large scale in the book; but all have been since removed and apparently but one saved—that of a satyr, whose oblique, leering eyes, animal ears and horns, and sardonic grin of self-satisfaction are inimitable, but with a certain dignity lent by an escallop shell rising like rays from the back (Fig. 25). This piece is now exhibited in the Museum. In Tijou's book all are supported by scrolls with acanthus and garlands, over a moulded stand and richly embossed Cloth of Estate, upheld by the two large lateral scrolls forming the main part of the screen below.

These, with their great acanthus leaves, none of which are now original, finish above in eagle-heads, holding between them a great garland of acanthus husks. No such sumptuous garden screen has since been attempted, and had Tijou's original embossed work been preserved it would have been among the finest extant works in wrought iron. The novelty and charm of Tijou's work were due to his masterly ability in designing and embossing masks, garlands, diapers, acanthus, etc., of sheet iron, and using them in discriminating profusion, combined, in the French manner, with smiths' work. In these respects no work in our country approaches Tijou's before or since (Fig. 26). No less satisfactory are the three pairs of gates closing the arched entrances to the garden front of the Palace. Thus sheltered, they appear in perfect preservation, but the stripping of the paint, which seemed likely to safeguard them for years to come, has revealed that practically all Tijou's personal work has been renewed at different times, under Government care, either too elaborately in cast brass, or too sketchily in sheet copper. The design remains, however, purely French, each gate divided into three panels in double frames connected diagonally at the angles. The upper is oblong, over a wide panelled lock-rail, with a square panel beneath. The centre gates are the richest, the upper panel with an unfilled oval, but designed for a bust or monogram ; the rest arabesqued with rich effect, due mainly to a judicious use of acanthus. The lock-rail is scrolled with eagle's heads and acanthus, and the lower panel is of close and stronger work, with less embossing. The side gates as executed are similar, but less rich in detail.

In his book of designs, Tijou indulges in flights of fancy, some almost incapable of execution, but, as stated in his brief preface, " all for the use of those that will worke Iron in Perfection and with Art." Designs of key-bows may have been for jewellers, but none like them exist. An eagle with a pierced disc may be intended for a mirror-stand and back, and there are other designs of more doubtful application, and almost prohibitive cost. Three pyramidal studies suggest designs for supports to his six vanes "finely wrought in leaves and scroll-work"—an unusual description, as elsewhere he usually calls even his richest productions "grotesk." It is doubtful whether all his gates, and certain that not all his staircase designs, were carried out. As Chatsworth was being built in 1687, Tijou may have been employed there soon after his arrival, and he is reputed to have made the railings for the south front, as well as balconies and the grand stair balustrade. The balusters are of lyre design, much as in his book, but without the connecting ornament. This particular form may be traced back to an engraving in a French design book of 1670, and it reappears enriched in Jean Marot's book before 1679. The staircase at Chatsworth has actually fine landing-panels of monogram and garter with coronet, and lesser panels on either side, not illustrated in the book; though an extremely rich balcony with the Cavendish stags'-heads and serpent crest is engraved in it. There were also gates by Tijou and palisading, probably for the west front, but now removed to the Lodge. The Burleigh gates near Stamford, with a full-page illustration in the book, were carried out as designed, the filling of the semicircle above them with almost the full richness of details. The wheatsheaf, known as "garb" in heraldry, with lion supporters on scrolls, almost fills the overthrow, but the gates below are of lighter construction with few moulded bars. Two pairs close the north and west entrances to the inner court.

During his thirteen years' sojourn subsequent to the publication of his book, Tijou must have undoubtedly worked for many other private clients. Among the most

important would be Mary, Duchess of Norfolk, and her
husband, Sir John Germain, a reputed half-brother to the
King, at Drayton House, Northants. The house is modelled
somewhat on the lines of Hampton Court, especially
the colonnades to the Courtyard, and the ironwork appears
to be contemporary, and either partly by Tijou or influenced
by his work. When the family papers are examined,
questions may be set at rest, but meantime some characteristics seem unmistakable. The principal gates to the
forecourt are between noble stone piers, recalling the famous
" Flowerpot " piers of Hampton Court. Railings to the
garden are simpler, with rather severe pilasters and pyramid
tops, crowned with three flame-tufts clearly developed from
Tijou leaf-clusters. A second grand façade with gates closes
the South Avenue, between fine stone piers with lead vases.
These are fixed in the centre of over one hundred feet of
railing, in thirteen bays, between pilasters, and surmounted
by pyramids of scrolls with water-leaves and groups of
flowers. The gates are quite unusual; the pyramid tops
forming part of them and opening with them, in place of
the usual fixed overthrow, are of scrolled acanthus and
laurel. Their design is almost reproduced in another pair
to the Orangery, both fifteen feet wide, inclusive of pilasters.
The railing pilasters in this case are of a Greek scroll repeating design, and the pyramids to the railing, of scrolls
and flowers, are quite original; and, if actually by Tijou,
show almost unexpected versatility, and deserve close
study. Yet another pair of gates to the East Avenue hangs
from pilasters entirely of scroll design, with pyramid tops,
in which acanthus is sparingly used. These gates of plain
bars have central decorative pilaster-like panels of repeating
scrolls, each with its high pyramid of scroll-work; ending
above in curious gourd-shaped finials of welded and twisted
iron. All have been detailed to inch scale and published

by the late Mr. Bailey Murphy, but, invaluable as these may be, the work itself deserves careful re-examination on account of the many original features introduced, since appropriated and brought into common use by smiths, whose claims to originality may thus be appraised. Moreover, leaf-work of importance to the designs may have perished. The work is from every point of view of quite peculiar interest. The balustrades of Drayton House are no less interesting than the gates, while suggesting Tijou influence in every line yet not appearing to be his handiwork. As the house is approached through the paved courtyard with its Hampton Court-like colonnades, a short flight of steps and wide landing lead to the entrance, with a very stately iron balustrade commencing with four-sided newels of scrolled iron panels under solid moulded pyramids finishing in gilt balls. The steps and their balusters are unusually wide and elaborate, and the half-landing panels of scrolls and acanthus are between pilasters of varied design, with little relation to each other, but are not seen together. A longer and better conceived balustrade from garden to drawing-room is of steps and a half-landing, with a much wider landing at the finish. The balusters are lyre design with acanthus, and the half-landing panel between pilasters under a scrolled border seems of Tijou design, but probably shorn of some of its embossed acanthus work. The principal landing panels are similar, but of enhanced dignity, with ducal coronets, and side panels in double frames, joined by diagonals at the angles, emphasised by four handsome pilasters. This repeats at right angles on a width of twelve feet, and all angles are marked by stout vertical standards with gilt bronze balls. A third balustrade is in the house, from the great hall to the King's dining-room, with a baluster of lyre design to each step. The landing has two panels and three pilasters, with a scrolled border. A Cloth of Estate enriches both panels and pilasters, which are perhaps less happy in design than the rest. The oak hand-rail, unless modern, is an early example.

The fine gates now at Cheveley Rectory, near Newmarket, removed from the Manor House opposite, were probably erected by a Duke of Rutland, and of the date and by the maker of those to Drayton House, but possibly by Warren. They are lofty, with unusually narrow piers of slender lyre filling and scrolled pinnacles, having open gourd-like points. The central pyramid of the overthrow, without definite outline, is mainly of branches of laurel, acanthus and tendrils, with a small mask and two groups of naturalistic flowers. The transom is of light repeating Greek scrolls, as at Hampton Court, and delicate buttress-scrolls lead down to the much lower wickets. At the Manor House is a fine balustrade of lyre design between standards, with scrolls and water-leaves in the manner of Tijou. At Exning, on the other side of Newmarket, is an exquisite gate, probably also by Warren, now opening from the garden to the road. It consists of a centre gate with arched top, with similar fixed panels, all between high and narrow pilasters. The filling of lyre design resembles those of Cheveley. The two centre pilasters end above in scrolls and large sprays of five lilies issuing from vases with drapery, and the side pilasters are capped with small buttress scrolls. The over-throws rise over the arches in pedestal form from the buttress - scrolls, richly worked with acanthus and laurel, to support a centre group of laurel and berries, with vases and lilies on either side. The gates are of vertical bars with dog-bars. Two other gates in the gardens with pilasters and overthrows are also of interest. The gates at Burley-on-the-Hill may be by Tijou, since they are stately, between pilasters of enriched verticals and scrolls ; and with scrolls in pairs below the lock-rail. They open under a high pyramid

overthrow, with coronet, monogram, Cloth of Estate, laurel
and acanthus.　Several of Kip's illustrations show gates,
like those to Lord Ossulstone's seat at Dawley, Middlesex,
with the dignity and proportions of Tijou's later works.

In Tijou's books are several rich and original designs
for staircase balustrades, possibly with a view to future
requirements at Hampton Court; but none, except that
at Chatsworth, are known to have been carried out. Wren,
as architect to Hampton Court Palace, had shown little
disposition to patronise Tijou's rich designs. Tijou was employed " task work " on rails to the King's and Queen's
back stairs, the King's privy stairs, the Princess's, Lord
Portland's, and others, including the " Beauty Staircase,"
the balusters merely a waved bar between loose scrolls,
or large ovals between single scrolls and verticals, arranged
without reference to the treads, and with plain iron hand-
rails. These appear to us now as quite unsuited to a Royal
Palace.　One of these is repeated in the balcony of an
entrance-hall, its coarse severity being mitigated by the
addition of two exceptionally rich supporting brackets
with eagle heads, looped with festoons and tied with
ribbons.

Not until three years later, 1699, was the King's stately
staircase balustrade requisitioned. The design is like nothing
in Tijou's book, nor in the French taste, and may thus have
been inspired by Wren. It is in panels two feet six inches
wide, mainly of vertical and bent bars supported by scrolls
and acanthus‑leaves, with richer pilasters at intervals.
The staircase hall is lofty and spacious, with Verrio's painted
ceiling, " the wall," as described in the building accounts,
" black and gold painted with armoury, like a wainscoat,"
the iron rails " carved and gilt."　A pleasing design in
Tijou's book is of a gate surmounted by an overthrow of
palms and the Royal Crown, and with bars scrolled at the

base and coupled above with acanthus, and four central scrolled panels below the lock-rail; this gate seems to have stood on the wings of the Palace to shut off the bowling-green and privy garden, such gates being represented in old engravings. Tijou, in this instance, seems to have borrowed a design by Briseville and developed by Jean Marot.

The balustrade, still intact, of fleur-de-lis design at the end of the Long Water, whenever put up, is clearly simplified from one in Tijou's book, though again originally adapted from a Briseville design elaborated by Jean Bérain. The Queen's early death was an irreparable loss to Tijou, for she had been a kind and invaluable patron, and he was perhaps, as a foreigner, left without other friends.

It is refreshing to turn from the mutilated work at Hampton Court to his hardly less important and still intact work in the Cathedral of St. Paul, then building by Wren and nearing completion. The detailed building accounts for each year are in the Cathedral library, and show that Tijou's first consignment was made in 1691, and delivered from Hampton Court. This consisted of the iron frames for the large ground-floor windows, charged at 6d. per pound, equal to about £48 each, exclusive of two vertical iron stays filled in with alternate circles and lozenges with rosettes charged as " grotesque." Tijou made the whole of the lower tier of windows, and all above those in the choir, transepts and west front, his deliveries of these extending to 1697. There are also large moulded brackets with acanthus scroll supports over the keystones, now not well seen, while some have been removed for decorative glazing, but their acanthus work now converted into electric brackets can still be seen in the crypt. The whole of the superb and important ironwork in the choir is by Tijou, mostly untouched and in perfect preservation. The gates to the apse followed in 1697, placed in position in time for the opening service of Sunday, December 5th, of that year. These sumptuous gates are treated as screens with a magnificent cresting of six large candelabra of scrolls, joined and elaborated by palm and acanthus, evidently based on French design. They are, unfortunately, not visible from the transepts. Another great work was the organ-screen, dismantled so late as 1860, which formerly stood between eight marble columns under the organ, forming the entrance to the choir, closed with gates and fixed panels, supplied by Tijou in 1696 for £442. The gates and side panels of this screen, in new frames and cresting, now form screens in the ambulatory (Fig. 27); parts of the old frieze are fixed over a new inner porch of the north entrance to the cathedral. The central gates on the north side present the only known instance of a design by Tijou seeming to be incomplete until both are closed. They contain finely embossed seated figures of the four Evangelists in medallions, with radiating ornament, while the former wicket - gates are of large ovals centring in seated figures over the lock-rail with vertical scroll panels beneath. That these are Tijou's own work seems evidenced by the fact that Hopton the joiner was paid " ffor gluing of Boards for Mr. Tijoue to draw ye iron screene upon." In 1698 Tijou delivered " two Desks for the Choristers seven feet long, each containing twenty-six panels, etc., and for four brackets for the same about twenty-four feet superficial by agreement £265." These have been dismantled, and only two of the small exquisitely embossed panels sixteen inches square are preserved, let into seats in the Chapel of St. Michael and St. George, the design being of two muscular terminal male figures falling away from a fluted vase, in one case with a mask above, and supported by scrolls and acanthus. The great gates to the Choir aisles are, perhaps, the most imposing, and were delivered in 1697. They

depend for effect chiefly on the grouping of enriched
verticals clustered in a dignified and stately fashion somewhat foreign to Tijou's usual manner. They open as a pair
between two large openwork Corinthian pilasters of close
work, beneath a frieze supporting three rich pyramidal
decorations. The gate panels are of four small Corinthian
pilasters with scrolled bases and an almost geometric
treatment between, and a simpler design below.

With the completion of the choir, Tijou's work in
St. Paul's practically concludes. The altar-rail now at the
entrance to the choir was not delivered till 1706, at a cost
of £260. It is rich and reposeful with acanthus, masks and
busts, but not otherwise striking, though useful to many later
smiths. In the same year he produced the balustrade to
the geometric stairs for 22s. 6d. per foot, total, £156 18s.
The balusters are like some at Hampton Court, though
more refined, and are reproduced in the Chapter House.
The ironwork begins on a half-landing with a peculiarly
fine architectural treatment of a central pyramid between
smaller four-way pilasters with obelisk finials, all of notable
design. Dog-gates close the stairway at this point. The rest
of Tijou's work for St. Paul's is external, the principal
being the remarkable gates closing the East and West approaches to the South entrance. These Tijou calls the
"great gates," "framed with strong iron and ornaments
and points on ye tops," for which he received £160 each
in 1697 and 1698. They are of immense weight, and though
ten feet wide and seven feet six inches high, open in one
piece. Moreover, the entire central panel of the three opens
as a wicket, the only instance known, and requiring very
massive construction and quite peculiar design. In 1705
Tijou accepted an order to make the great chain to strengthen
the dome, another engineering feat, with £20 extra for
extraordinary work in the joints. His only other works

are the window grilles. The earliest seem to be in 1694, for two little windows £20, and £40 for the rails to staircases on each side of the great West door. There are two designs of grilles for the North and South porches, each used twice, altogether eight grilles. One of these is a design of ovals with scrolls penetrated by radiating bars, and the other of more intricate scrollwork centring on a circle and lozenge. There are also three pairs of grilles in the West porch, one with semicircular top with a fleur-de-lis as centre, and a geometric treatment of circles, scrolls, etc., below, in which stamped rosettes and other peculiarities of forging may be noticed. Another fills two small semi-circular openings with a fan-shaped design springing from an exceedingly fine mask ; and the third is circular, of four geometric scrolls supporting a moulded circle of radiating ornament, made in 1694. All are exceptionally beautiful. His final work connected with the Cathedral was in 1711–12 for a railing round the newly erected statue of Queen Anne, on the space in front of the West portico. It was as seen in Bowle's view, 1753, of perfectly plain design, and petticoated or bulging outward in a curve away from the base. Tijou may have waited on expecting perhaps the great order for rails and gates to enclose the twenty acres of church and churchyard, which Wren probably meant him to have, and for which upwards of £11,000 was eventually paid. He departed, leaving his wife to collect £385 13s. 9d., which was paid to her on November 1st, 1712, " by virtue of a letter of attorney from my husband, John Tijou," signed "Anne Tijou."

That Tijou returned to Paris and died poor and broken-hearted is probable, since the original copper plates of his book were re-published there. At first these still bore his name, but afterwards that of Louis Fordrin, a designing craftsman. Whether his wife actually followed him as instructed is, like everything else, purely matter of conjecture. That Tijou was a great artist and superb craftsman cannot be denied, and he was probably reticent, making few friends. That the figure in redingote and cravat, with hair tied behind, adjusting a leaf to an iron bar by aid of a very brawny smith, is a portrait of Tijou, by his son-in-law Laguerre, can hardly be doubted. The original pencil drawing of this fortunately exists, and shows a strongly marked face with heavy moustache but forbidding expression. The magnificent scale of his book of designs, and the employment on it of so many well-known artists of repute, was an ambitious undertaking, even for a Frenchman in the days of the Grand Monarque, and wholly unprecedented in England. His signature shows that he was an educated gentleman, and might possibly provide an insight as to character. In the earlier accounts of work at St. Paul's he is styled " Monsieur Tijoue," but this slid into John Tijoue, J. Tijoue, and Tijoue, always the final " e," due to a strange terminal flourish. His daughter married Laguerre, the fashionable decorative artist, and they and the Tijous were parishioners of St. Martin's, then embracing Piccadilly and Soho. He does not appear as rated, perhaps as an alien, and thus neither the actual place of his abode, nor sites of his works can be traced. Notwithstanding the striking originality and high quality of his work, his name is mentioned but once in contemporary literature, and then only casually by Vertue to quote a cynical remark on Laguerre. That Wren should never have alluded to him, though speaking so much of another foreigner, Grinling Gibbons, is strange, and equally so that Evelyn, who watched his fine works erecting in St. Paul's, or Pepys, who let very little pass his vigilant notice, should have no word to say. Wren, as architect, may naturally have resented his interference at Hampton Court, and having so far exhibited little

G

partiality for ironwork, kept his interior clear of Tijou's designs for several years after the Queen's death.

Neither is the name mentioned in France or Holland, not even by his compatriot Daniel Marot, who actually caused the removal of his great garden screen at Hampton Court to an inferior position. That he left relatives in London seems clear, for a Mrs. Anne Tijou was buried in St. Martin's in 1708, and a male Tijou in 1709, while people of his name continued to carry on business in the parish, and the name of a T. Tijou is engraved under two rather poor designs of balustrades of half a century later. Some of his name still living claim to be his descendants.

As to the Huntington Shaw controversy, little need now be said. Tijou was not personally a smith, and lived in London, and may or may not have required a foreman of works at Hampton Court. Shaw described himself in his own will as a blacksmith, born in Nottingham, June 26th, 1660, and would thus, if he ever met Tijou, have been about thirty years of age. Nothing whatever is known of him or what he did at Hampton Court, but when he removed to London, in 1700, he lived in Frances Street, now 17, Air Street, and was rated at 8s. by the Parish of St. James until 1707, when the books are missing. He died October 20th, 1710, aged 51, having made his will three days before, bequeathing his bills, bonds, and book debts to Mary his wife. Benjamin Jackson, Queen Anne's Master Mason at Hampton Court, was sole executor, and erected a fantastic monument to his memory, twelve feet high, against the exterior wall of Hampton Church.

The inscription as given by Lyson, and also transcribed by Sir Hans Soane, ended, " he was an artist in his way." The monument disappeared in the rebuilding of 1833, but the tablet appeared inside with the vacant space left to record the widow, filled up by the addition in different writing of the words, " he designed and executed the ornamental ironwork at Hampton Court Palace." There is no shadow of ground for the statement. Mr. Garraway Rice, F.S.A., took great pains to trace the matter out, having accidentally secured the iron monogram of the railing originally in front of the monument.*

Thus Tijou is vindicated, but, apart from his book of designs and works, he remains a mysterious figure, coming from nowhere, but apparently with a wife and band of relatives, and disappearing to die obscurely in Paris. His good fortune was to meet the Queen and gain her confidence, but his evil genius brought him in collision at Hampton Court with the dangerous Talman. He left no memoirs or even letters, accounts, or diaries, and his addresses are unknown.

* R. Garraway Rice, F.S.A.—*Notes on Huntington Shaw, blacksmith, in* " *The Archæological Journal,*" *Vol. LII., pp. 158—172.*

CHAP. VIII.—THE FOLLOWERS OF TIJOU.

OF the English smiths who were influenced by Tijou, and whose work is treated of in the following chapter, William Edney of Bristol, Robert Davies of Wrexham, and Robert Bakewell of Derby adhered more closely than any others to his style, making free use of sheet metal for embossing acanthus, draperies, masks and the like. Other men, such as Thomas Robinson, George Buncker, Partridge and Warren, produced work in which the construction lines are more prominent, and are without the decorative additions which almost smother Tijou's work; the result is greater dignity and stateliness. By degrees the English smith, whose energies had been stirred by the work of Tijou, developed a national style of ironwork characterised by stability and gracefulness of design as well as by technical skill, a style which recognised at once the possibilities as well as the limitations of the metal.

Among English smiths contemporary with Tijou whose names are now known, THOMAS ROBINSON is the most important. He first appears in the building accounts of St. Paul's in 1697, as the maker of the rail to the Morning Prayer Chapel, the precursor, perhaps, of all the London area-railings with decorative spikes. In this the horizontal has a deep moulding on each side over the verticals, while above it rise graceful leaf-shaped spikes on bulbous supports, eleven inches high, made hollow and with shorter forged moulded spikes between. The standards are heavy, and have finely shaped turned finials. The work was difficult, but skilfully executed, at the price of 10d. per lb. £104 9s. od. He is next heard of, 1699–1703, sharing the upper windows of the nave at 5d. per lb., though Thomas Coalburn had 6d. Robinson also made a rail in 1704 for the arch in the north-west side of the dome, of very graceful panelled design, at 10d., and no doubt the three others which match it, and in 1706 a rail for the Consistory Chapel, probably that now in the crypt, with curious moulded spikes and geometric panels. In 1708 he made the balcony round the dome for £300, and in 1711 one thousand feet of chain for suspending lamps, four iron fences and gates for the West portico, and two for the North, and for the Chapter House, including a model for a fire-back, which he carved. If the balcony made in 1708 was that to the " Whispering " Gallery, it is a massive work, with panels of crossed swords at intervals, garlanded with laurel, between geometric pilasters and heavy standards surmounted by balls; the rest being simpler, of moulded ovals between scrolls, with acanthus. The Earl of Lanesborough is said to have erected the Western Gallery at his own expense; the balcony is by Robinson. It crosses the west end, in three panels, two with bishop's mitres and one with a censer, the rest of the rail being more simply panelled in scrolls in pairs, with two acanthus leaves and a drapery. It continues over the cornice for about fifty feet on the north and south sides, but has since been continued along the entire cornice by the gift of Mr. G. Somers Clarke. It seems likely that some of the railing work at Hampton Court may have been entrusted to Robinson by Wren, but the accounts do not as yet seem to have been exhaustively examined. At this time, 1711, Robinson was completing the very fine garden screen at New College, Oxford, one hundred and twenty feet long, Mr. Ayliffe * stating it to be " by that ingenious artist, Mr. Thomas Robinson, at Hyde Park Corner." Miss Fiennes

---

* John Ayliffe — *The antient and present State of the University of Oxford. 1714.*

notices it as in the middle of the first quadrangle, where there was a stone statue of William of Wykeham " railed in with iron gates." The garden was " then new making, a large basin of water in the middle and walles, mazes, and roundabouts." The scheme of a single richly worked gate, between fixed panels and piers, with wide over-sailing transom panel and pyramid top, unusually lofty for the width, produces a fine and stately perpendicular effect, never surpassed. It is enhanced by the wide expanse of high railings on each side, with their curving plan and fine pilasters with pyramid tops, and continuous and well-balanced and varied cresting over the railings, which are of perpendicular bars with handsome dog-bars. For the gate Robinson has adopted a setting-out somewhat like Tijou's, especially in the four-centred panel below the lock-rail, radiating from a circle. There is no embossed work about the gate, but two small cast masks appear in the upper part ; and above these the bishop's arms in a garter, and motto beneath a mask, also two acanthus leaves, with the mitre for apex (Fig. 28).

The gates to Trinity College, Oxford, were placed by Wren at the end of a new avenue of limes in 1713, between three-sided or triple piers, intended for wicket gates. The ironwork seems to be by Robinson, and is of vertical bars with scrolled lock-rail and handsome dog-bars and fringe. The pilasters have finely designed lyre panels set between four standards, and a finial of four foliated scrolls set diagonally to a centre. A wide transom of scroll-work connects these over the gates, supporting a great richly worked pyramid of scrolls, tendrils, laurel and acanthus, bearing the College arms, and a mask and crest, with lesser pyramids over the pilasters. Fixed panels above on either side dip in semicircles, making room for a scroll design. These were at the far end of the gardens, shut off from the College quadrangle by a second high screen with gates, now it appears removed to form the College entrance, less fine than originally designed and much restored. Defoe, speaking of the gardens, mentions that "at the entrance and end of the walk are very noble Iron gates, which leave a prospect open to the whole East side of the College."

The fine gates of New College tempt us to attribute many others of no less good proportion and design to Robinson, especially as the firm existed for over a century, but there are few decided peculiarities about this work to constitute an unmistakable sign manual.

Another contemporary of Tijou was PARTRIDGE, a London smith, who made the handsome cloister grilles and gates under the Library of Trinity College, Cambridge, built by Wren. They are about seven feet wide, and consist of double and single gates and fixed panels, divided almost equally by horizontal bands of ornament between vertical bars, fringed above and below with C-scrolls and wave-points. The design for the panel in the wicket gates occurs in a Wren drawing, preserved in All Souls' College, and those for the double gates were no doubt also developed by Wren, who may have instructed Partridge. The garden side is closed by grilles of inch bars bent over at the top to form concentric semicircles, one under the other, and permitting the passage of radiating bars which traverse and hold them together; the spandrels are filled with massive scrolls. The stair balustrade to each step is a stout twisted bar with moulded cap, branching immediately under the handrail into a cluster of water-leaves, tendrils, and supporting scrolls. It is included in Tijou's book, but may have been inspired by Wren a couple of years or so before the book was published in 1693. The hand-rail is of one-and-half-inch iron, handsomely moulded, and the newel a four-sided hollow pillar. The railing in front of the Master's Lodge at Trinity looks contemporary, comprising handsome baluster panels alternating with plain bars, to which scrolls are welded in pairs. Fine lamp supports with buttressed scrolls of similar date are on the garden balustrade.

Another smith employed at Cambridge at a slightly later date is WARREN, who produced the three gates at Clare College. The simplest is that to the " Backs," of three panels of almost plain vertical bars divided by two horizontals, the centre opening as a gate, the others fixed under a restrained overthrow of scrolls and arms; their horizontals fringed with scrolls or spikes, with short open-work dog-bars below. These were produced in 1714. The front gate is similar, but more richly treated, and all three panels open under the transom and overthrow. The central panel under a semicircle of radiating bars and scrolls is used as a wicket; the fixed transom of two strong bars with scroll border between, supporting three pyramids of scrolls, the centre bearing the College arms and acanthus foliage. The Bridge gates of Clare are the most important, and were put up by Warren in 1714, between four pilasters treated as the gates, except that below the lock-rail their panels are scrolled: the dog-bars, lock-rails and transom are richly worked in horizontal bands of decoration beneath pyramids with somewhat " Chinese Chippendale " silhouette. Above the central gate is a shield between great branches of laurel, or perhaps mistletoe, loaded with foliage and fruit, with acanthus tendrils and scrolls and finials above. The high railings at either side and at right angles have spikes to render them unclimbable. For these Warren received £326 11s. 6d. Thanks for our knowledge of Warren are due to Mr. J. R. Wardale, of Clare College. Warren was a fine craftsman and skilful designer, and no doubt also executed the four gates to St. John's and those to Jesus College, of

about the same date. Gates to Bulwick Hall, near Oundle,
are evidently by the same smith; their flat pilasters are wide
and filled with weak scroll-work without cross-ties, a
defect compensated by very strong stay-bars, richly scrolled.
In these the pyramid is without definite outline, chiefly
remarkable for two stiff fan-shaped branches of seven
lanceolate leaves, tendrils and mask, and a monogram in
the garter and crest.

The name of another thoroughly accomplished smith,
GEORGE BUNCKER, is known as that of the maker of the
gates to the Inner Court of Dulwich College in 1728. Though
a single wicket might have sufficed, the space between the
fixed pilasters is filled by a pair of lyre panels, which open
as gates. On either side are panels of duplicated lyres,
which give dignity, the whole under a handsome overthrow, with central pyramid bearing the arms of the
donor, Edward Alleyne, his crest as apex, and motto,
" God's Gift," inscribed on a semicircle below. These
are enclosed and supported by scrolls and buttresses with
water - leaves, tendrils, and spikes, and lesser pyramids on
each side of similar design. A lighter pair of gates in
Dulwich village is also by Buncker, and in good proportion, without either arms or crest.

Much of the work in and about the Metropolis might
safely be ascribed to these known London smiths, but nothing is more certain than that the names of very many others,
still unknown, may yet be discovered. Moreover, the fact
that so much of the work was in public, open to the in-
spection of all the competitors, who could assimilate any
novel departures or combinations, make all pronouncements
unsafe in the absence of actual records. Robinson must
be the author of a vast amount of fine work in and about
the Metropolis, for he had a long career; and Buncker, a
name still favourably known in smithcraft, must also have

been responsible for much; while the peculiarities of Warren should make it possible to recognise his work; but the fact remains that there are still large groups of London work as to the authorship of which we remain absolutely ignorant.

With smiths working in the Shires, on the other hand, it is wholly different. Their reputation and the demand for their work spread and increased, without plagiarism or mutual influence. Each developed in his own way, and an expert may speak quite definitely about them, and, when once recognised, their characteristics are not easily forgotten. The smiths who divided the work in the Shires, hitherto discovered, are PARIS of Warwick, EDNEY of Bristol, BAKE-WELL of Derby, and DAVIES of Wrexham, and they almost parcelled out the West and Midland Counties between them. None are known so far to have held any similar monopolies in the Northern or Eastern Counties.

After the partial destruction of St. Mary's Church, Warwick, by fire in 1694, PARIS was employed to put a new railing to the Earl of Leicester's tomb in 1716. Though inappropriate and of scrolled panels of unnecessary variety, it is otherwise interesting, with moulded rail, gadrooned vases and wrought flame-tufts. The choir railings (1706) and gates comprise multipetalled roses and acanthus, all of good and original design. There were also two mace-rests, presented by Queen Anne—one with her monogram. The gates and railings to St. John's Hospital are by the same smith, and they repeat his very narrow panels with tulip drops and flowers terminating scrolls; but the overthrow has, unfortunately, become hidden or destroyed by ivy. Perhaps, the remarkable screen from Frome Church, in the Museum, No. 1092–1875 (Fig. 29), may be an earlier work by Paris, the perching birds and laurel branches proceeding from gadroon vases being novelties such as he indulged in.

The crest, shield, and minute monogram in this relate to the
Lords of the Manor. A gate with three feathers in the overthrow, the crest of the Mompessons, in the Close of Salisbury
Cathedral, also appears possibly to be by Paris, and the
extremely narrow panels of the pilasters of another plain
gate on the south side also point to him as the smith.
Perhaps the peculiar characteristics of the screen to the
Christopher Eyre tomb in St. Thomas's Church at Salisbury,
made in 1730, connect it with Paris. It was removed from
the outside to the inside of the church in 1893, probably a
re-transfer. The screen appears to be a three-quarter scale
model of a pair of gates with overthrow, piers and wickets,
with armorial bearings. Over the centre is one of Paris's
gadrooned vases with flame-tuft of wrought iron, and on each
side are flowers in his manner. It is no less difficult to resist
assigning to Paris the gates of St. Mary's porch at Oxford,
which stand so completely alone in design and workmanship. Their overthrow is treated horizontally, raised on a
base of large running scrolls and a narrower border above ;
it centres in a flask-shaped handled vase, in silhouette, of
bent bars like shell ornament ; from each handle sags a
twisted bar in rope fashion, clustered with bunches of tulips,
roses, and other flowers; beneath are the University arms
on tasselled drapery. The gates above the scrolled lock-rail
are of bars joined in pairs by moulded collars and scrolled
at the base, and the dog-bars are arrow-pointed, each with
pairs of scrolls and of leaves. The filling of the gate piers
is as those of Warwick, with higher pyramids, surmounted
by a bunch of flowers, and they widen at the base in an unusual manner (Fig. 30). The gates to the President's garden,
Magdalen College, comprise a large circle with monogram and
crest between two sharply bent moulded bars, with a branch
of holly on each side, some light scroll-work and a festoon
of acanthus-husks for overthrow. The gates beneath are of

vertical bars with a fringe above of acanthus-husks, and scrolls at the base in place of dog-bars. The pilasters have moulded caps supporting large vases in silhouette of bent iron bars. The great work by Paris may probably have been that at Rugby Hall, Alcester, in Warwickshire, a seat of the Marquis of Hertford. An old engraving shows this well provided with gates, the central pair in the screen to a forecourt, with bowed front, having a rich overthrow and two fixed panels between stone piers, with about two hundred feet of railing. Four additional gates are shown in the front wall, and six on the garden side. Photographs show that three pair of gates were in existence until about forty years ago, all between high stone piers, with moulded caps and fluted vases. One of these bore in the pyramid of scrolls forming the overthrow the monogram and coronet, in a small oval, of Popham Conway, the first lord, supported by boys and enrichments of acanthus and rosettes. Of the overthrow, only the base of scrolls and two cornucopiæ with remains of flowers, fruit, scrolls and rosettes exist; but the transom, also of scrolls and flowers, and the entire gates, though patched, were perfect and most characteristic of Paris's work. The third pair was quite complete, the overthrow higher than usual in Paris's work, centring in a small circle with monogram, supported for the most part by scrolls ending in his peculiar rosettes, branches of laurel, husks, and acanthus. These seem to have been removed, but a pair still exists to the kitchen garden, apparently much restored, with the arms, supporters, coronet, crest, and motto of Lord Conway fully displayed in the overthrow. In Paris's designs the unexpected happens: some want of proportion in the parts, some confusion in design, poverty or over-richness, or unneeded variety. They are naïve, but often charming, and where remaining should be rescued. He appears not to have worked in London,

but over a wide field elsewhere. A small example,
perhaps one of his best, was turned out of St. Mary's,
Taunton, over thirty years since. It comprised
chancel-gates and side panels of rich scrolled work
with acanthus, more free of mannerism than usual,
and in perfect preservation, and is now at Hunter-
combe Manor, Taplow, the property of Lady Eleanor Boyle.
Another fine example of gates exists in Solihull Church,
Castle Bromwich, between fixed panels, under an overthrow,
perhaps an early and characteristic work of this fine craftsman.

WILLIAM EDNEY follows Tijou's tradition more closely
than his contemporaries, and is chiefly known for his
fine work in the Bristol churches. Payments are made
to him in the St. Mary Redcliffe accounts, in 1710, for iron
gates to the chancel, and two pair of gates in the side aisles.
In 1722 three iron gates were fixed in the chancel of
St. Thomas's, Bristol, demolished in 1789, when possibly they
were transferred to St. Mary's, which now possesses more
of his ironwork than that set forth in the accounts. His
work has, unfortunately, been rearranged during more than
one " restoration." The chancel-gates are of rich design, based
on those of the garden entrance to Hampton Court Palace,
but with the added effect of no less rich pilasters with
acanthus caps and high pyramid overthrows, and a still
higher step-gable pyramid between, with the arms
of Bristol. The gable idea, which comes from Antwerp, is
worked out in scrolls and acanthus, with a gadrooned vase
and wrought flame-tuft on each of the six steps. The main
panels comprise small masks and much acanthus, with
many of the scroll-ends made to droop vertically in moulded
points, with graceful effect. Below the lock-rail the panels
are four-centred, as in Tijou designs. The rest of the ironwork in this church is less rich, more original, yet hardly

so entirely satisfactory. The fine baptistery screen in St. Nicholas' Church consists of a pair of gates six feet high, loaded with acanthus, and with very delicately worked monograms. The piers, eight feet high, are no. less richly worked, surmounted by draped vases (Fig. 31). In this church is also a mace-rest with eagles' heads and monogram. The fine gate and screens of the Temple Church, also in Bristol, were supplied in 1726, these and the mace-stand bearing the monogram of George I. It is all less florid and more geometric than the older work (Fig. 32). Earlier in date are the gates to Tredegar Park, near Newport, Monmouth, between four-sided piers and wickets in which no part is plain, and the whole effect is perhaps even rather marred by richness of detail. The pyramid overthrow, centring in a mask supported on a scroll base with smaller mask, has a regal effect. The gates are of repeating panels of scroll-work with less acanthus, and thus of greater transparency; but the general effect is marred by the prodigious weight and size of the four-sided acanthus capitals, and their somewhat poor pyramid finials. The filling of the pillars is close and intricate, and, as duplicated, not easy to decipher: they are repeated on a smaller scale for the wickets, which are less laboured than the rest of the design. The gates given by Lord Gage to Tewkesbury Abbey bear initials and a baron's coronet, and must consequently have been made before he became a viscount. He became Member for the town in 1734. The overthrow centres in an oval with monogram, between dragons' heads holding acanthus-branches, but above it, in rather ugly fashion, is an unsupported shield embossed to commemorate the step in the peerage. The gates are of coupled bars, from which the acanthus caps have been removed, with scrolled panels between; and the four-sided piers have capitals as ponderous as those of Tredegar but also stripped of acanthus and other enrichments in a restoration.

ROBERT DAVIES, born near Wrexham, is another smith
with great ideas and capacity, who made a big mark. He
was apparently trained by his father, a smith, who died in
1702, and joined by one or more brothers. Sir Richard
Myddleton, of Chirk Castle, was an early patron, and the
fine gates commenced for him in 1715 were not completely
finished or paid for until 1721. The extraordinary variety
and splendid quality of much of the work expended on these
show a determination to excel, apart from pecuniary results.
The bill for these and the long stretch of handsome railing
on each side has been found by Mr. Myddleton, the present
owner; the amount is £190 1s. 6d. The central pair of
gates, of wonderful elaboration, are designed to a much
smaller scale than the rest of the screen, and are consequently
overweighted by the immensity of the framing. The introduction of eagles' heads as terminals to acanthus scrolls,
over four panels below, shows that Robert Davies was
conversant with Tijou's pattern-book, which must, by
that time, have been in the hands of all enterprising and
ambitious smiths. Each part of these gates is of elaborate
richness, and appears to be designed as carefully as a
diploma piece. The whole together, however, suggests that
no complete design was ever drawn to scale, and in no other
instance have such elaborate and costly details been introduced into every part of a work of such magnitude. Thus,
all the verticals in the panels of the great four-sided piers
are balustraded in the Spanish Renaissance manner, and
there are eighty of them. Minor details, such as the small
drops, spikes and dog-bars, have complicated weldings
and mouldings, which might require from a day to two days
each to produce, and quite subordinate laurel sprays needed
thirty or forty welds. It is easy to realise how it happened
that some four or five years slipped away in their production.
Even the railings have crestings of unusual richness, and

each dog-bar in them has a pair of leaves, a pair of scrolls,
a pair of. barbs, three ball swellings, and carefully shaped
javelin points : 160 of them remain, and possibly originally
there were 200. The gates are surmounted by a frieze of
acanthus and a large central rosette. In each main panel
is a central pedestal of verticals supporting a Medusa-
like mask on drapery, filled in with and buttressed by scrolls,
with eagle-head finials, acanthus and flowers. The lower
part is even more crowded with detail. The fixed pilasters,
transom and overthrow seem worked out to almost twice
the scale of the gates, with scrolls bearing foliage, tulips
and other flowers, yet as to design hardly meriting description. The work, however, is elaborate with strange conceits,
and, as smithing, remarkably fine. The piers are four-sided,
unnecessarily massive, with colossal stone caps. Beyond are
two fixed panels in the guise of wickets, better as to proportion.
The whole was produced to enclose a paved court in front
of the castle, seen in Buck's view of 1765. In 1770 the gates
were removed to the Park entrance, but since restored, in
1888, with the railing, to their original position (Fig. 33).

In 1718 a chancel-screen by Davies, of plain and twisted
bars, was erected in Wrexham Church. The gates are of
scrolls with cherub heads, and curve downward towards the
centre, supporting a remarkable group of vine with foliage,
fruit and tendrils; while on each side the rails are crested with
a curious *espalier* group, possibly of cherry, with branches
of tulip flowers in plate iron. This work is clearly suggested
by a plaster ceiling at Emral, near by where Davies was
working. So curious a screen was condemned as "Grecian"
in 1833, but saved by Sir W. W. Wynn. Gates were put up
at Wrexham Church in 1720, and also to Ruthin Church, and
in 1738 to Oswestry Church, and Davies probably also made
those to Malpas Church. These works are not, however, very
remarkable, nor their proportions admirable. The gates

to Emral are more interesting, of light work and pictur-
esque in design, with two vertical panels of scrolls
centring in a trefoil of acanthus, and between scrolled
pilasters. The high overthrow with large shield of arms
is loosely constructed of scrolls and laurel branches. The
gates are celebrated as appearing in "Washington Irving's
Sketch Book," by Caldecott. A small pair of gates
closing a bridge over the Colbrook are also well designed,
of repeating light scroll tracery and leaves of easy-
flowing lines, with stronger pyramidal cresting and four-
way pilasters. There is also a charming dog-gate on the
staircase, and a beautiful openwork vane over the stables,
bearing the date 1733. Other gates, at Carden Hall, and
Abbey House, Shrewsbury, are clearly by Davies; but
whether we may go further, and safely attribute the magnificent screens of Leeswood, near Mold, Eaton Hall, and
Newnham Paddox to him is still debatable. All these are
by the same hand, and exhibit a power of design and sense
of proportion and fitness lacking in Davies' earlier work ; and
yet the man who made the Chirk Castle gates could have,
and is the most likely man to have made these also, and in
that case Davies would be among the foremost of British
smiths. No other smith is known to whom they can be
attributed so safely.

ROBERT BAKEWELL'S work is more distinctly based on
the French than that of any other known contemporary
smith, but he never directly borrows from Tijou, who for a
brief space was his contemporary. He made free use of
acanthus and rosettes, his drawing and handling of these
being excellent. His leaves, cut out and worked on the
anvil, are consequently stouter, crisper, and more flowing
and twisted than those produced by embossing on pitch.
His early source of inspiration is clearly traceable to the
great fore-court screen of the French Embassy, Powis

H

House, in Great Ormond Street. This was built in the time of William III., burnt down, and immediately rebuilt in 1712 at the cost of Louis XIV. The screen comprised large entrance gates, and a handsome stretch of railings on each side, divided by pilasters into bays with scrolled borders and fine pyramid tops, all of rich and excellent design. These, and the no less handsome area railings nearer the house, and probably a fine staircase balustrade within, would easily have provided all the ideas, details, and lessons in technique necessary to the intelligent, observant, and highly skilled craftsman Bakewell. He may have worked here, but, if not, might at least have studied the work from the street at leisure. He made constant use of it, and having a sense of balance and proportion which rarely failed, his designs are always successful, if not strikingly original. Also when he found a satisfactory *motif*, he used it again and again, and hence his work can rarely be mistaken for that of his contemporaries. After some first attempts he hardly ever produced masks, which require to be embossed. In these he wholly lacked expression and power, and thus usually substituted escallop-shells. A characteristic, so constantly used as to do duty for a sign manual, is the filling, wherever possible, of long narrow panels and spaces by a single closely waved bar, which simplified work and gave it the Bakewell *cachet*. He is first heard of in 1707, at Melbourne, near Derby, working on the construction of a domed porch to a garden house, suggested no doubt by the arbours of treillage of similar form then popular in England and France, but this one of iron has remained unique (Fig. 34). It was made for the Rt. Hon. Thomas Coke, and commenced in August, 1707 ; but illness supervened, and the bill was not rendered till April, 1711 : £120, as agreed, with a few extras. It consists of a dome, about nine feet in diameter, resting on three walls of masonry. The construction is of bar iron, the spaces filled with scrolls and branches of laurel and oak alternating, surmounted by a central cage-like cupola of scrolls under a ball. Over the entrance is a feeble mask and Cloth of Estate with scrolls, resting on a lintel of ovals, upheld by two narrow pilaster-like panels in advance of the rest, to form a door-frame. The façade is completed by well-designed panels, between pilasters of vertical bars carried above the lintel, and finishing in clusters of vases and a crescent on each side. Bakewell intended to set up a shop in Derby on completion of his work at Melbourne, but Elizabeth Coke reports that he was " so miserably poor " that he could not remove till he had some money (April 8, 1711). No further dated specimen is known until the beautiful gate for John Lombe's silk mill at Derby was set up, which bears his initials in monogram in the overthrow, and date 1717. The plain gates open under an arch of scrolls on handsome pilasters, beneath a pediment of shell-like design, framed and buttressed by scrolls and acanthus. Much of his work remains in Derby, but his greatest work produced between 1723 and 1725, as detailed in Dr. Cox's fine quarto volume on the Church of All Saints,* is the fine screen which formerly extended across the east end, enclosing the Cavendish Chapel with the monument of Bess of Hardwick (Fig. 35). Needless restorations have left the central portion to form the chancel-screen, while a part is removed to enclose the Cavendish and other monuments on the north wall. The screen in its entirety was a magnificent work, consisting of a large central arch and overthrow, with pilasters. The gates, which were removed in 1878, were between four-sided piers, and surmounted by a lofty overthrow of scrolls and acanthus, culminating in the Royal Arms and supporters over a Cloth of Estate. These piers,

*J. Charles·Cox and W. H. St. John Hope—*The Chronicles of the Collegiate Church or free Chapel of All Saints, Derby.*

H2

characteristic of Bakewell's work, are of beautifully designed panels, between heavy standards, their handsome moulded caps surmounted by panels of short vertical bars buttressed by scrolls, supporting large characteristic vase-like finials of open-work. The screen is of vertical bars, coupled in pairs with acanthus capitals. The former side entrances, now at right angles and without gates, comprise open-work pilasters with spandrels and frieze supporting a pediment of shell design. A frieze of ovolos and short panels with delicately waved centre-bars between, united the entire screen, and over this were pyramids of ovals, scrolls and acanthus. The whole almost rivalled in dignity that by Tijou in St. Paul's Cathedral, and the cost, with a pair of rather plain gates to the churchyard, was but £388 10s. The altar-rail, Communion-table, and rail to Thomas Chambers' monument in this church are also by Bakewell. His other works in Derby are the gates to the Baptist Chapel, and to premises now occupied by the Gas Company; and elsewhere in the county, to Longford Hall, Mill Hill House, Norton, Etwall Hall on the road to Melbourne, Tissington Hall, and Ashbourne Church. None of these, however, probably equalled in grandeur the fine range of gates made for Wingerworth, subsequently removed to Savile House, in Leicester Square, and thence, after a long interval, to Penshurst, where I was privileged to fix them, still bearing the substituted monograms of Frederick Prince of Wales, but otherwise shorn of much of their pristine magnificence. One of Bakewell's most successful attempts is still at Okeover Hall, where the gates somewhat resemble those formerly at Devonshire House, Piccadilly, even the stone piers and vases being similar. Another pair in the grounds at Okeover, with two pair of four-sided wrought pilasters for smaller gates, bears a shield of arms. In the county and beyond its limits are the gates to the Guildhall at Worcester, 1723,

Cannock Chase, Staffordshire, Newark, Nottingham, Wolverhampton, etc. The balustrade at Melbourne repeats the altar - rail of All Saints, Derby, and another similar is in Ravenstone Church. Other fine works are the domed font-cover in St. Werburgh's, Derby, railings in Hicker's Row, and the fine choir-screen, in attempted Gothic taste, in Manchester Cathedral. Stair balustrades at Melbourne are repeated at Okeover, St. Helen's House, Derby, and 160, High Street, Hull.

In Carlisle an unknown local smith has left five or six simple but bold overthrows to gates, and some brackets, the scrolls sometimes intercrossing with an occasional tulip, multipetalled rose, looping tendrils, or barbed arrow-points, on waved shafts. There is no old ironwork in this Cathedral, but Miss Fiennes records that it was entered over a bridge through " double gates which are iron gates and lined with a case of doors of thick timber " ; thus a fortified place. So capable a man with original ideas would hardly confine himself to a single border town, and elsewhere we meet with his work in York, Lancaster and Nottingham. The balustrades to High Head Castle, though later, may be by him. He, no doubt, had opportunities of seeing other smiths' contemporary work. The choir-aisle gates in York Minster re-introduce well-known earlier features, such as the cock's head and acanthus of Tijou, and the clusters of peculiarly lanceolate laurel-leaves and berries, and the barbed arrow-points, on long intertwining stems. They replaced wooden gates, and are illustrated by Mr. Bailey Murphy, and ascribed by him to Dean Finch, in the eighteenth century. They comprise four vertical panels, much restrained, while the arch above is filled with radiating scroll-ended bars with welded leaves. There are apparently brackets, pilasters, and railings by this smith in Micklegate and elsewhere in York. In Lancaster, the gates in Market Street, St. Leonard's

Gate, and Castle Hill House, as well as those removed from
the church and broken up, seem all to be by him, as are
also the fine pair at Wavertree Hall, near Liverpool, those at
Old Park, Winchmore Hill, and a simple pair at Hornby.
Certain of the work at Nottingham, including the gates
from Colwick Hall and at Newark, may safely be attributed
to the same source, with the strangely proportioned gates
from Quenby Hall at Leicester. Several gates are met with
in Northumberland and Durham with wide transoms; the
design is a central circle, from which proceed three diverging
bars right and left, as the width may permit, ending
in scrolls, with which the inter-spaces are also filled.
Above is a low scrolled pyramid with a second circle,
usually with monogram. The gates below comprise
panels of rather peculiar lyre design, scrolled lock-rails
and barbed dog-bars, and the gate-piers where present are
massively framed, each of six heavy standards with scrolled
panel and finials. Examples of single central gates at
Tanfield Manor, near Durham, and Jesmond Hall,
Newcastle-on-Tyne, are similarly designed between corresponding pilasters. A pair of gates to the Rectory of
Chester-le-Street only varies in having the panels of the
pilasters built into stone frames and caps, an almost
unique example. This same as yet unknown smith
was apparently taken South to execute the ironwork at
Knole, near Sevenoaks, or his work was copied there, his
wide transoms, of most striking design, being reproduced
unchanged, but with a pyramid considerably modified. In
these the monogram and coronet of the Earl of Middlesex
and Dorset form the central feature of the design. The
pilasters, too, with similar lyre design, are taller and more
attenuated, while between these and the gates are additional
fixed panels of scroll design twice repeated. Neither these
nor the garden gates by the same hand were produced until

after 1716. It seems possible that the Sandringham gates, with the monogram of one of the Henly Hasle family and date 1724, though vastly simpler, and the gates at Lynn, dated 1714, may also be by him.

In the near west a local smith conversant with the prevalent style of design seems to have produced some important gates. This may well have been John Davis, who carved his name on the mechanism to the bell of the Curfew Tower at Windsor, which he repaired. The gate at Hadley House, Windsor, and those to Baylis House, Slough, are locally attributed to him. Gates at Mapledurham Church, Welford Park, and New Compton, Berks, Chilton House, Bucks, and a small wicket at Cookham, appear to be somewhat heavier in execution, and are interesting, though not quite up to the London standard. Two pair at Aldermaston, and two at West Woodhay, Berks, and those at Miserden, near Cirencester, are obviously by London men. Scrolled gates of purely local work are in the garden of Penshurst, and at Chiddingfold, and the stair balustrade of Cobham Hall, in Kent, seems also the product of a local estate smith.

CHAP. IX.—EIGHTEENTH CENTURY WORK IN GREATER LONDON AND ELSEWHERE.

THE very large group of gates in Greater London is, on the whole, of better design than those met with elsewhere in England. An interesting point is whether they were supplied by some industrial firms in London or by local smiths. Though now practically united, these hamlets or villages were, less than a century ago, entirely separate. A certain group of gates distinguished by peculiarly graceful lines and details, usually has the overthrows drooping from a centre with tent-like outline. Most of the existing examples are found about Stratford and its neighbourhood, but others are also found at Stoke Newington, Hampstead, Highgate, Tooting, and Chiswick. The craftsmanship is excellent, but the work late—1720 to 1730. A second group, of earlier and sturdier work, is mainly distinguished by its wide pilasters with rich filling, usually associated with plain gates, but often with a central panel matching the pilasters, the overthrows relatively unimportant or absent. Examples are as wide apart as Iver, West Drayton, Chiswick, Twickenham, Chelsea, Stoke Newington, Tottenham, and Enfield. The distribution of these and other groups seems to prove that the more important and scholarly-designed gates were produced in London, while others of less skilful work were made locally. The writer possesses drawings and photographs of over two hundred, not all still existing. Over one hundred of these are to the west of London, about half on the north side of the Thames, and the rest at Clapham, and between Putney and Richmond. Between forty and fifty are to the north, chiefly at Stoke Newington, Hampstead, and Highgate; none between Hampstead and Acton. On the Essex side, over fifty have been sketched or photographed. On the south are fewer, some thirty in all, chiefly in the neighbourhood of Carshalton, where there seems to have been an exceedingly able local smith.

Of gates actually made for their present positions in the London of the eighteenth century but few exist. The most interesting are the Bridewell gates, preserved in the offices in Bridge Street, Blackfriars, and those at the foot of the stairs of the Headquarters of the Honourable Artillery Company in Finsbury. Both are early examples. The Inns of Court have preserved a fine pair, with the initials T.W.I.C. in the overthrow, and the date 1723; though somewhat severe, they are handsome and preserve the best traditions. A single gate to the Temple Gardens has wide, fixed panels on each side, with pilasters and overthrow, bearing the winged horse, a roundel with crest, and the date 1730. This and the preceding are by the same smith. There are old railings to some of the Temple " Walks," those to Nos. 4 and 5, King's Bench, attributed to Wren, having interesting spikes and panels. A handsome gate and railing, of which no vestige remains, enclosed the parterre in Fountain Court. The original railings, with both the old lampholders and others more modern, are preserved to Lincoln's Inn Fields, where many of the houses also retain their old railings. A fine stair-balustrade is at No. 66, in the northwest corner, with its handrail and lamp-holder; No. 33 preserves its panelled rail. Serjeants' and several other Inns have kept their railings. A handsome gate and screen to the gardens of Furnival's Inn has disappeared. Important gates and screens to Lindsay House, Chelsea, as to Newcastle House, have also vanished.

Most of the gates and railings to London hospitals have also disappeared, such as the stately iron gates and gateway presented to St. Thomas's Hospital by Mr. Guy in 1707, at a cost of £3000 ; and those to Bedlam. The only iron screen to a London hospital forecourt remaining is at Guy's, in Southwark ; but, if old pictures are to be trusted, the whole must have been restored and considerably altered. It is, however, still handsome and satisfactory. The most remarkable group of ironwork in London is that to the Children's Hospital in Great Ormond Street, put up when the residence of Lord Thurlow, in 1704 : it comprises handsome railings with panels, gates and lamp-holders, and a fine balustrade to the steps leading to the garden at the back.

When churches had to be kept open for ventilation, iron gates were fixed in the porches to prevent unauthorised ingress, and several of these remain. The specially fine gate and the window-grilles to St. Paul's Church, Covent Garden, severe in design, but with heads of classic fan tracery, are, fortunately, intact ; they date from 1727. Most of the old church porch gates still in London are of plain spiked bars and dog-bars, with lock-rails filled with circles ; as to Bow Church and the Church of St. Anne and St. Agnes. That to St. Dunstan's-in-the-East is further enriched with a frieze of cherubs. Those to St. Sepulchre's are probably not original, and the pair in the not distant porch of Christ Church, Newgate Street, are unusually decorative. An oval in the overthrow contains the letters L.S.C. repeated and entwined, while above and below the scrolled lock - rails of the gates are panels of scroll-ended bars, with a central vertical bar bearing diverging pointed leaves. An interesting overthrow to a gate closes a passage to the hospital not far distant. Unfortunately, the much finer gates and railing to St. Andrew's, Holborn, were destroyed in the making of the Viaduct. Nearly all the churches of London have had their burial grounds abolished or reduced, and the older railings replaced by new.

Within the City the Livery Halls and public buildings were generally secured by iron gates and rails, hardly any remaining except those to the Bank of England, by George Sampson, erected in 1734. The gates are severe and massive, yet appropriate. A gate in St. Catherine's Court, near the Tower, though almost entirely re-made, can be studied in the remains preserved and hanging on the wall hard by. The finials are cast. Another, under a semicircular arch, to the Marine Society, founded in 1756, is of good design, with "M.S." in the grille. Except near Stationers' Hall, there seems no longer to be any old ironwork existing *in situ* associated with City Livery Companies. Of such works Halton writes, in 1708 : "Those especially about half a mile in compass round the Royal Exchange, particularly eastward therefrom, are so numerous and magnificent, with Courts, Offices and all other necessary apartments . . . and noble gates and frontispieces of some towards the street . . . that it would require too much room to give the names and situations." A careful drawing of the massive iron gates and inner court of No. 102, Leadenhall Street, the last of these fine old mansions to be swept away, was made by J. W. Archer, and published in 1875. Numbers of decorative window-grilles existed in the City till about the middle of the last century ; of several of these illustrations are preserved. Westminster is even worse off. Almost the last pair of gates, those to the Dacre Alms Houses, disappeared when these were pulled down to make way for St. James's Court. Only the gate to the Little Cloister of the Abbey now remains, besides the later screen and gates to the Horse Guards. Many interesting railings with lamp-holders and ramps, and some with panels, existed in Smith Square, Downing Street, Great George Street, and about Whitehall and Spring Gardens, but relatively few remain. Nearly all the fine rails enclosing churches, such as existed in

St. Mary-le-Strand, St. Giles', and St. Martin's-in-the-
Fields, are either abolished, or replaced with new.
Even the railings round the Abbey and St. Margaret's
have undergone many changes and " improvements." Fortunately, interiors have suffered less, altar-rails and pulpit-
balusters having in many cases been spared. As compen-
sation, the superb ironwork of St. Paul's Cathedral is for the
most part intact, and the finest in the world; while the
relatively little remaining in the Abbey is still deeply
interesting.

Railings, though incidentally mentioned, have not, so
far, been specially described. When associated with splendid
entrance-gates they assume the dignity of screens, in which
decorative effect becomes the chief consideration. They
are lofty, with high scrolled pyramids and overthrows,
and enriched pilasters, with scrolled horizontal borders
fringed with leaf-points, and finishing below in finely worked
arrow-pointed dog-bars. Thus, they rival the entrance-gates
themselves in stateliness. Several such have been described,
but none now exist in London, where they were once perhaps
far from rare. We have only now to deal with railings such
as are constantly seen in our older streets and London
squares and gardens. These are invariably formed of
seven-eighth or one-inch rectangular bars, set vertically
in a kerb, and passing through one, or rarely two horizontals.
They finish in a spike from five to nine inches high, shaped
by help of a tool, with a base swelling slightly and tapering
upwards to an obliquely cut point. They often alternate
with others barbed, or with leaves or scrolls in pairs, and
are then spaced about four inches apart. Curved bars as
stays give additional support when the lengths are long, and
these necessitate stouter bars, called standards, to support
them, as we have seen in the tomb-rails, whence the idea of
vertical bars for rails has been derived   The standards at

first finished in bolder spikes, clustered scrolls, or bulbs, but later were more generally surmounted by cast-iron vases. Tijou was the first to use such finials rather freely, especially in his work for St. Paul's, but his were either of repoussé iron or of cast bronze. Probably there was no thought of using cast iron for railing work until the cast Sussex iron railings were fixed round St. Paul's, in 1714. Wren for some reason objected to these, and they lay in dock for four years after delivery in London. He says, " As to the iron fence, it is so remarkable and so fresh in memory and by whose influence and importunity it was wrested from me, and the doing of it carried in a way which I venture to say will ever be condemned." Perhaps Wren had some stately idea, to be carried out by Tijou, who left England when the cast iron was actually delivered, in 1710. It was invoiced from Lamberhurst; but part was sublet to Mayfield ; and it completely environed the Cathedral, including seven gates, and weighing 200 tons. The price charged, 8d. per lb., was enormous, as other castings were sold for 2d. ; the cost amounted to the then almost fabulous total of £11,000. Probably for this reason no other cast railing is known to have been made in the Weald. Each vertical is a massive baluster, finishing above in a four-sided spike, and at intervals there are ponderous standards of the same finish and design. Rather close imitations are at the Oxford Schools, and Cambridge Senate House, but these have a wrought bar between each pair of cast balusters. A lighter example is at St. Leonard's, Shoreditch, put up by Dance in 1740 ; and cast standards of similar design alternate with wrought rails round the statue of Henry VI. at Eton. Meantime we must notice that Gibbs had introduced cast iron balusters in his altar-rails, as in St. Martin's Church, about 1726, with success. Cast standards of varied and enriched design, but generally balustered, soon became

familiar in London railings, the rest of the bars being plain
and wrought. The designs may best be studied in the squares
and streets north and south of Piccadilly, also immediately
north of Oxford Street, and as far east as Bloomsbury.
They are considerably varied, sometimes taking a columnar
form. In nearly all cases they will be found surmounted by
cast iron vases, the most popular being a flask-shape with
turned foot, spherical body, tapering neck and ball-stopper.
This is rivalled by an Italian gadrooned vase on a foot,
both forms introduced by Tijou. There are other forms of
vase, but more rarely seen until the Adam style prevailed.
Besides the vase, a ball finial is sometimes used, or a
mace-head design with fluted umbrella-shaped top over a
spreading base, as seen in Whitehall Yard and Gwydyr
House, 1796; the pine cone and the acorn, the former at
Kensington Palace and near the Horse Guards, and the latter
in Cheyne Row and Serjeants' Inn; they occur together in
Upper Brook Street. A curious cast design of two moulded
scrolls meeting under a flask-like knob became a cheap
substitute for wrought finials on gates, to be seen on those
of St. John's College, Cambridge, and elsewhere ; and these
occasionally replace vases on railings, as in Serjeants' Inn
and on the north side of Clapham Common.

The spaces in the London squares and parks were at
first left open, or enclosed only by wooden posts and rails.
Lincoln's Inn was the first to be enclosed, by the existing
railing, in 1735. Soho was similarly treated in 1748, at a
cost of £950. Leicester and Golden Squares followed.
St. James's may have been enclosed when the pedestal for the
William III. statue was erected in 1732, or not long after.
Grosvenor Square followed in 1753, and Portman in 1764,
the old railing being now, however, replaced by a modern.
In all these there were standards and dog-bars ; in some
a few taller standards of wrought iron were adapted to

carry lamps. These and brackets fixed to the walls were the only forms of street lighting in the eighteenth century. The lamps were glass bowls with iron or copper wind-proof tops, which were lifted off and replaced to receive the small oil reservoir and burner ready lighted. An iron ring attached to the standard or bracket held the glass. Original standards to hold them remain at Lincoln's Inn, together with a later and more enriched arching form. Some of the houses in St. James's Square erected stone obelisks for the lamp-rings, and a few others exist in neighbouring streets, and until lately lent interest to enclosures of Hanover Square and to the west side of Westminster Abbey. Wall-brackets with rings for the lamps are severely plain, as in Sackville Street and elsewhere. In other cases they were much ornamented. It soon became a fashion to provide the door-ways of mansions with a pair of wrought iron standards of rich design to take lamps, generally rising from the area-rails. Otherwise, a ring for a lamp is suspended between two arching brackets over the steps. Both kinds may be extremely decorative, and even fine specimens of smithing and design. The handsomest are in Berkeley, Grosvenor, and Portman Squares, building when the fashion was at its zenith (Fig. 36). Among the happiest are those in which the duplicated and foliated scrolls cross in a French manner, akin to Chippendale's florid style, as at No. 45, Berkeley Square. There are modern reproductions, but originals of this same design recur in Portman Square and Chesterfield Street. Lamp-holders to the front doors occur in most of the streets of Mayfair and a few south of Piccadilly, and also in the squares immediately north of Oxford Street, and east as far as Bloomsbury. Others exist about Whitehall and Westminster, but many in Great George Street and Spring Gardens have now been swept away. Similar but less rich lamp-holders are, or were until lately, to be seen in

Edinburgh, Bristol and Bath. Hogarth shows this type of
lamp in two of his pictures, and De Loutherbourg in one.
Large and high wrought iron posts to take several lamps were
erected at Hyde Park Corner and in a few other public places.
They lasted till the end of the century, when the more modern
rectangular glazed lanterns with a door took their place.
Some arching supports of wrought iron to lamps in Bristol,
Bath, and Edinburgh are excellent. Wrought iron brackets
were produced for many purposes, as for hanging lanterns,
especially over the wells of staircases. There are several of
rich design in the old Ball Room of the Mansion House,
from which are hung tent-shaped glass chandeliers of the
Regency period ; another very fine specimen is on the
staircase. They are based on Tijou designs with eagle-heads
and acanthus. There may be many such in existence to be
met with by chance. Brackets were also fixed against walls
or pilasters with welded rings for the glass basin lamps,
several remaining in the Admiralty Court Yard, built by
Ripley, dating probably from 1726. No doubt some of the
many brackets in the Victoria and Albert Museum may
have been to hang lamps from, but far more frequently
they were used as supports to inn signs, shelves, hoods over
doorways, landings, and so on (Figs. 37, 38). Probably the
richest ever designed for the latter purpose are those by
Tijou at Hampton Court. Others may be found in
various parts of England, and from the similarity and
general excellence of design were perhaps merchantable
articles sent from London.

Iron handrails and stair-balusters were probably first
needed for short flights of steps leading from one level to
another in public buildings, such as Westminster Hall, and
cathedrals like Canterbury ; but were not perfectly developed
until the introduction of well staircases. The most obvious
support for the handrail is one, or a pair, of vertical bars

to each step. Any design or decoration must conform to echeloned conditions, since the steps are one above another, their rail forming an angle of about thirty degrees, with each baluster limited to about one foot horizontal and about three feet in height. With two perpendiculars on each step, the interspace may be filled with ovals, scrolls, and so on, later examples in considerable variety being met with. The simple vertical can be forged into a lozenge, oval, or heart-shape, or otherwise decorated, as with lateral scrolls and foliage. The " S "-scroll is well adapted to supporting the handrail and filling the space below; such scrolls riveted to one or two short bars to let into the stone, were largely used. Enrichments from a simple pair of welded scrolls, or water-leaves, with, perhaps, a ball or some tooling between, are further elaborated with tendrils, clusters of bay - leaves, rosettes, etc., until the rich, baluster effect is attained. With such, advantage was generally taken of landings to intro-duce even more effective treatments and variations. Some of the lost houses in Great George Street, as Nos. 11, 28, and 29, afforded appropriate and interesting examples, which may possibly be preserved by the London County Council. At times the original forms were almost lost in a profusion of embossed acanthus-leaves, as in the Blooms-bury and Berkeley Square examples, and the old War Office in Pall Mall. The illustration of the balustrade to the garden steps of the Manor House, Wandsworth, in the Spring Gar-dens Sketch Book (Vol. VIII., Pl. 57), presents a simple re-arrangement for the landing. The pleasing example in the Museum formed the altar-rail of Uxbridge Church. The most favoured design, however, was based on the lyre, introduced from France, and susceptible of practically endless variation, even becoming unrecognisable in some of its extremes. The lyre may be classic, as treated by the Brothers Adam. It may form the whole design, resting

I

on a base ; or only the central feature, when a subordinate design may form a base assuming some importance.

When perpendicular lines are introduced, these are short or interrupted, unless central ; but pairs of verticals uniting in a stilted arch above are of frequent occurrence. Sometimes later balusters verge towards rich rococo. Examples may be found in most of the good streets and squares built in the second half of the eighteenth century. Many designs are in the Chippendale taste, and some few introduce the traditional embossed acanthus, either as leaves or rosettes. Among notable examples is the fine balustrade to Wren's staircase in Marlborough House, and that to 15, George Street, Hanover Square, both early examples, considerably enriched with embossed acanthus, dating from about 1710. The pulpit-rail in St. John's, Westminster, is on similar lines without repoussé-work. The Museum possesses some examples (Fig. 39) ; there are upwards of twenty, all different, at Inwood, near Templecombe, the seat of Lady Theodora Guest. In all, the writer has sketches of about seventy different treatments. Many more must exist in private houses, in town and country. Designs for Communion-rails are frequently similar to those for staircases, but these are generally introduced as foils for more elaborate central panels. Richer and continuous forms of stair-balustrades reached us from France. After using rather straggling and indifferent designs for the secondary staircases at Hampton Court, Tijou produced the really rich and fine design for that to the magnificent King's staircase of black marble. Though he did not remain in England to execute the corresponding Queen's staircase, for it was not produced till 1731, nearly twenty years after his departure, the balustrade design is based on one in his book. This seems also to have been rearranged for the altar-rails of University

College Chapel and St. Alphege, Greenwich. The staircase by Wren in Kensington Palace is in a series of
repeating panels much wider than high, as in Tijou's
designs, but the absence of embossed work is against it
being his, and the presence of a light mahogany hand-
rail of French section would, unless a later addition,
make it probable that it was not erected till 1730.

The groups remaining to be described cannot as yet be
directly attributed either to architect or to known smiths, but
their identity may yet be discovered through building accounts,
where such exist. One of the most striking of existing groups
is represented by the gates formerly at Devonshire House,
and now set up on the opposite side of Piccadilly. They
were purchased by the Duke of Devonshire when Heathfield House was pulled down in 1837, and ultimately brought
to Piccadilly (Fig. 40). They repeat two of the peculiar
square quatrefoil panels and wide lock-rail of the old gates
formerly at Buckingham House, St. James's Park, and to
gates at Sunbury, the latter now refixed at Culford Hall,
Bury St. Edmunds, in which such panels are twice
repeated on a larger scale in the wickets, and also
below the overthrow, with the acanthus and gadroon
border used by Tijou in the Hampton Court gates.
The central mask of this border and those on the
pilasters are of cast iron, these castings being frequently
used on other gates, premonitory of decadence in smithcraft. The coat of arms, now a conspicuous feature of the
overthrow, was inserted by the Duke; it is supported by buttress scrolls with laurel leaves grouped in threes, and berries
on long, waving and intertwining stems. The lofty pilasters
of lyre design terminate in somewhat over-sailing pyramids
with acanthus, repeated on a broader scale over the wickets.
Gates of identical design, but differing proportions, are at
Clandon, near Guildford, high railings and pilasters taking

the place of wickets ; while acanthus caps to the piers are somewhat heavy, a fault corrected in the railing pilasters with pyramid tops. The design again repeats in the fine gates and screen to Kirkleatham Hospital, near Redcar, without wickets, but with lofty railings with scrolled cresting and pilasters of lyre design. These were certainly fixed after 1714, and are believed to have been purchased and brought from elsewhere, but probably not till the chapel was built and buildings altered in 1742. This hospital also has an extraordinarily rich wrought iron altar-rail and stair balustrade. Gates to Kirkleatham Church and Hall are simpler, and evidently later than those to the hospital. Somewhat similar gates with lyre pilasters, and overthrow with acanthus, cocks' heads and drapery are at Kirklees Hall, Brighouse.

The entrance gates now at Aldermaston are said to have been removed from Midgham House, Berks, both having been the property of Lord Stawel. They are richly worked and apparently by the same smith. A single lofty gate with lyre pilasters and gadroon border under the overthrow has been removed from Stoke Newington to Inwood, Templecombe, comprising a monogram of " T.G. " interlaced in a circle, and including characteristic laurel branches, and three cast masks ; laurel garlands also hang in front of the acanthus caps of the pilasters. Another example, even simpler, remains at the " Queen's House," No. 16, Cheyne Walk, Chelsea. This gate is plain, with semicircular head, and duplicated lyres one over the other, form the pilasters. The monogram " R.C.," interlaced, set in the midst of buttress scrolls, with laurel leaves rising high above it, is that of Richard Chapman, who built the house in 1717. The railing heads each bear a pair of leaves, scrolls below, and button points ; and the lyre pilasters are high with scrolled finials. At the back of the house are boldly designed balconies formed of the owner's initials, " R.C.," interlacing with pilasters having cast masks, too frequently introduced by the smith in these later works. There are also graceful balusters of lyre design to the front steps. Others belonging to this group, no less interesting, are the spiral stairs and gates, of extraordinary magnificence, made for Horseheath, Cambridgeshire, built by John Bromley, who died in 1718, leaving instructions to his executors " to finish the building he had begun to erect at Horseheath Hill according to his first design, and also the garden walls, *iron gates* and all things necessary." He anticipated delay, and they may not have been finished till after 1720. The Hon. Henry Bromley, then M.P. for Cambridgeshire, presented them to Trinity College in 1733 (Fig. 41). Their design is most dignified, rising from the railing pilasters in step-like gradations to the wickets and central gates, the superb overthrow reaching a height of twenty feet. Each rise is marked by buttress brackets, and increasingly large and lofty pilasters of lyre design, which, with the overthrow, form a rich foil to the somewhat plain gates. The overthrow rises from a low scrolled transom and buttress brackets, taking an almost semi-circular outline, and now comprising the College arms over a Cloth of Estate. Above is one of the small cast masks already mentioned, the Bromley crest no longer forming the appropriate apex, over a second Cloth of Estate. The main piers are richly worked and capped by obelisks outlined over scrolls. All forgings are unusually bold, the ultimate convolutions of the scrolls produced to widths of from six to seven and a-half inches. All these just described are of relatively late date, and their chief characteristic is a peculiarly lanceolate type of laurel leaf grouped rather formally, either in pairs, or more frequently in threes, with single berries, on slender round stems which are waved and

gracefully intertwine. These, though not even remotely
in Tijou's manner, have unfortunately been taken as the
correct type in the repairs by the Office of Works
to the Tijou screens of Hampton Court, to which they now
form a rather conspicuous sort of cresting, falsifying the
date by some twenty years or more. These screens were
probably the *magnum opus* of Tijou, and nothing com-
parable to them is seen later. Their work should on no
account have been tampered with.

Two pairs of lofty and large gates in the garden and
the church at Hatfield are of plain verticals with single lock-
rail, and barbed dog-bars, forming a most complete contrast
to the work at Hampton Court. Their overthrows are confined to buttress scrolls and finial, supporting a low stepped
rectangle over the saltire arrows and coronet of the Cecils.
The piers are high, severe in design, capped by rich pagoda-
shaped pyramids with acanthus. These are reproduced on
a lesser scale in a wicket gate and pilasters to the garden of
Ball's Park, Herts, which bear the monogram of Edward
Harrison, who succeeded his father in 1725. The balcony
over the front entrance is also interesting. Richer ironwork
at the Priory, Reigate, is clearly by the same smith. The
double gates each bear a semi-circular head two-thirds their
width, the spandrels filled with scrolls beneath a transom
scrolled and with pyramid top. Pilasters repeat the Hatfield
design with pagoda top, but enriched. The wickets are plain,
but also with handsome scrolled overthrow. In the Priory
grounds are the screen from the forecourt, and garden gates.
Railing repeating the design of the screen, removed to the
churchyard, now surrounds a grave appearing to date from
1703. Gates formerly in West Street have disappeared,
but the railing in Church Street yet remains, with the fine
panel and overthrow to the wicket. Gates taken to Little-
cott, in Wilts, are by the same smith, as well as a far richer

pair at Hall Place, Bexley. Others to the Grange, Farnham, date from at least as far back as 1704, and like a second pair in Castle Street, repeating Tijou pilasters and low pyramid tops with acanthus, are no doubt early works by the same smith. The approach to Wotton, near Aylesbury, is singularly impressive from its magnificent gates and railings of wrought ironwork, between massive stone piers and vases. These consist of high centre gates with wide scrolled transoms and low overthrows, comprising a central shield of arms and palm leaves, environed and supported by scrolls with acanthus leaves and rosettes, laurels, finials, and clusters of short twists welded to a point. The gates are between stone piers, but on either side are fixed panels little less lofty, with pilasters, and wickets opening beneath a semi-circle of radiating scrolled bars and leaves. Above the transom are five pyramids of scrolls with acanthus, surmounted alternately by clustered twists and cast vases. The transom and lock-rail carry through, and are of parallel bars with scrolled ends. A gate matching these is in the garden, and over a small wicket the rich and beautiful relic of Canons, Edgware. The Chelsea cast mask recurs in the transom over the central gates of Wotton as it does again in the centre of Lord Bathurst's gates in Ciren-cester. In these the overthrow rests on a semi-circle of radiating bars and ornaments, with stepped buttress scrolls and acanthus ; a coronet and garter having been added early in the last century.

Belton House, near Grantham, possesses another fine array of ironwork,* partly that made for Sir John Brownlow, who became Viscount Tyrconnel in 1718, the rest of which is at Hough, a property which came into the family pos-session in 1743 ; he shortly after took the gates to Belton. The main entrance gates, between stone piers, have each

* Thoroughly illustrated by the late Mr. Bailey Murphy.

a narrow, decorated vertical panel and scrolled lock-rail,
and open beneath a wide transom of horizontal bars and
scrolls, and a moderately high pyramid overthrow, with
the arms, motto and supporters of the Tyrconnels ;
in the railings on each side pilasters bear the family crests.
The courtyard gates are similar, the overthrow somewhat
less rich, and the arms, without the quarterings, are confined
to an oval. The piers of the gates, with acanthus caps, are
sixteen feet high, almost repeating those of the Horseheath
gates at Trinity College. Railings on each side have pyramid
crests and scrolled pilasters, repeating four times, which
may possibly be later additions. A single gate from Hough,
with overthrow, is over fourteen feet high, and forms a
stately centre to a wide stretch of railings with lyre pilasters,
shutting off the Wilderness. A panel over the gate carrying
up the vertical bars below it gives the gate a lofty aspect.
The centre of the pyramid thus appears rectangular, with
buttressing scrolls, shield of arms, and coronet, and
beneath, the cast mask. A wicket in the garden between
lyre pilasters supporting coronets, depends mainly for
decorative effect on bunches of laurel leaves in threes, and
berries on long, intertwining stems. The work can be no earlier
than 1718. Two garden gates with monograms in circles,
and buttress scrolls and acanthus forming the overthrows,
are at Cranham Hall, Upminster. Another early group,
distinct and restrained as to ornament, comprises but few
examples, all near London. They are large and stately,
set between heavy pilasters, mostly with panels of lyre
design. The gates are of plain bars with scrolled lock-rail and
dog-bars. One near Battersea Church, and another to Abney
House, Stoke Newington, dating from about 1695, only
open centrally, like wickets. The overthrows are of good
and simple design, and there are pyramid caps to the piers,
either scrolled or of pagoda-like outline. Another at Fenton

House, Hampstead, opens as a pair of gates, and a fourth is at West Drayton House. A pair of gates formerly at Lindsay House, Chelsea, now reconstructed and removed to 5, Cheyne Walk, somewhat resembles those of West Drayton.

Another small group, in striking contrast, is represented by the gate to the Little Cloister of Westminster Abbey. The upper bar curves inversely to the cloister arch above, and admits a scrolled top with acanthus and barbed arrow points characteristic of the group. In this diverging scrolls are welded to the verticals, and a central lyre panel passes vertically from top to base without the usual interrupting lock-rail. A pair of gates of similar design at West Green House, Winchfield, has the central panel to each gate, but crossed by corresponding panels forming a lock-rail. It ends above in a pointed arch, with a probably modern scroll cresting. A similar gate is at Solihull, originally finishing with a semi-circular head, its overthrow replaced by a cresting resembling that at Winchfield, but filling the spandrels and holding a tablet commemorating the gift of the gate to the church in 1746. A small but charming wicket at Inwood, Templecombe, with scrolled centre panel and pyramid top, is between graceful pilasters with lyre panels. Another, with similar but less pleasing pilasters, is at Whitchurch ; and a third at Marlow, in which the gate design is repeated in fixed panels on either side. Two introduce barbed arrow points amidst the scrolls, and peculiar abrupt recurves with button points marking the passage of some verticals into scroll ends. The touch is light and graceful, the designs appropriate and pleasing, considerably varied in detail. All the examples here cited are unusually small. Surrey, especially on the London side, is rich in gates, and amongst some of the finest in England are the great screen and gates of Carshalton Park, now removed. These were made for Thomas Scawen before his knighthood in 1714.

The screen is 113 feet in length, and consists of lofty and massive piers architecturally designed, with solid moulded caps and bases terminating in crown-like open tops of eight bars, curved and converging to support the solid turned finial. These tops are four feet six inches high and two feet in diameter. Their simple dignity and architectural fitness, great relative height and crown-like finials, recall the now dismantled railings to the river bastion at Hampton Court, reminiscent of Tijou, but on a vaster scale. The transom, over a foot deep, is formed of eight horizontal panels of scroll-ended bars relieved by cast masks of fauns, and is penetrated by the two gate pilasters of lyre panels, finishing above in arching tops and cast vases. The central gates open under a rich pyramid of scrolls with the Scawen arms and crest, with buttresses and gadrooned base. The rivets and some slight remains show that this was considerably enriched with acanthus-leaves when in its original state. Beneath are the gates, plain with scrolled lock-rail and barbed dog-bars, and beyond these the wickets, of the same height and design, but with lesser pyramids, six feet high. Between are the great lyre pilasters, finishing above in arching tops and vases, while again beyond is a bay of lofty railings on each side, their top horizontals deflected to receive scrolls and a pyramid, and still farther beyond these, high pilasters of four standards and lyre panel, with finial of scrolls and vases. The screen terminates with much longer bays of identical railing, with more simple pyramid breaks in the centre, and stone piers carved and bearing lead figures of heroic size. A pleasing example of a gate on a smaller scale is in the Museum ; in this the applied acanthus foliage characteristic of work of the earlier part of the century entirely disappears. It probably dates from about 1750 (Fig. 42).

Our pre-Reformation churches were swept by the Puritans

almost clear of metal-work, except memorial brasses, and practically nothing was introduced in its place until Communion-rails were authorised, in 1662, " as before the Rebellion." They had been ordered by Archbishop Laud to be placed close to the altar, and are so placed by Wren. They were of wood, and probably one of the first in iron is the fine example by Tijou, in St. Paul's, the only one known to be by him, as well as the only original existing in any English cathedral or large Abbey church. The leading architects of Queen Anne, and until 1740, including Wren, Hawksmoor, Archer, Gibbs, Flitcroft, Dance, and James, commonly provided handsome wrought - iron altar - rails in their lesser churches.

It was different with the mace and sword holders, which were not part of the fixtures, but gifts from private donors. As City churches in every ward were liable sooner or later to be visited in state by a Lord Mayor, mace and sword rests are found in most of the civic churches. They are conspicuous objects, richly worked and surmounted by a gilt Royal crown, painted blue or green picked out with gold, and comprising three or four shields, the Royal Arms in garter, civic arms, the donor's and those of his livery, sometimes including a Lord Mayor's cap. All these arms are usually painted on copper, but may be on wood, variously shaped, either ovals or shields. Occasionally they bear a date, as in St. Dunstan's-in-the-West, 1743, but most are evidently much earlier, while a few possibly date from the close of the century. Some are extremely simple, as in St. Bride's, Fleet Street ; St. Bartholomew's the Great ; St. Botolph's, Aldgate ; and St. Peter-le-Poer. Among the richer examples are those in St. Helen's, Bishopsgate ; St. Swithin's, 1710 ; St. Katherine Cree ; St. Andrew Undershaft ; St. Margaret Pattens, 1723 ; and especially the several fine groups of four in St. Edmund King and Martyr ;

St. Olave's, Hart Street; Allhallows, Barking; and St. Mary-at-Hill. One of these has the shields enamelled on porcelain, and the supporters modelled in full relief. In some cases, where there are more than three, some have been rescued from some neighbouring church pulled down. Essentials are the small cup or escallop near the base, and the clip above to hold the mace or sword. A pair of such rests, uncrowned, are placed on each side of the entrance to St. Mary Abchurch. Some even of the most important churches—Bow Church; St. Stephen, Walbrook; St. Michael, Cornhill; St. Mary Woolnoth; St. Alban, Wood Street; St. Sepulchre; St. Andrew, Holborn; as well as several others — are now without either sword or mace stands. Other cities with mayors and corporations found citizens to provide the churches with mace and sword rests, such as may be seen in Norwich, Yarmouth, Lynn, Salisbury, Gloucester, Chester and Bristol. The Museum possesses a possibly unique example of a mace and sword stand supporting a light iron arch connecting them, from Newcastle-under-Lyme. That of Exeter is now in the Town Hall. The Leeds mace-holder consists of a rectangular bracket filled in with an arabesque of pierced sheet iron. Those of Norwich are single uprights with leaves or scrolls to bear the shields, varying in date from that in the church of St. Simon and St. Jude, dated 1590, to that of St. Peter Mancroft, dated 1809. The Gloucester mace-rest is in St. Nicholas' Church, with the Royal Arms and G.R., and there is a curious cresting to the organ-loft, apparently designed from sword-rests. Norwich and Exeter seem to have been the only cathedrals having special rests for the civic sword or mace. As works of art, none of the provincial examples seem of especial interest. In the Exeter example the mace is placed horizontally, and there are no shields. A richer example in the Museum is crowned.

Hat-pegs were combined with the mace-rest in St. Mary's Chapel in St. Magnus the Martyr, but this was hardly dignified, and they were more often on a separate and sometimes elaborately worked stand. Fine examples are in St. James's, Garlickhithe ; Allhallows, Lombard Street ; and St. Michael's, Paternoster Royal, all remarkably different in design. Those in St. James's are spikes with a few twisted leaves supported by " G "-scrolls. That figured by Birch,* with pegs in all for sixteen hats, is of three uprights, the centre bearing leaves and arrow-points, supporting a horizontal, over which is another central group of tulip scrolls and arrow-points, with a scroll and similar work on either side. This early example may date from 1682. The St. Michael's example, taking sixteen hats, is a much richer design of scrolls and leaves and of somewhat later date. The pair in Allhallows, also early, is the most severe in design, of twisted bars interrupted by welded scrolls. There is a good specimen in the Museum, and one of some antiquity, in form of a bracket, in old Chelsea Church. An early eighteenth-century example is in St. Martin's, Droitwich. Large quantities must have been removed, as they are now somewhat rare objects.

When bulbous brass chandeliers came into fashion, they were frequently given to churches, singly or in pairs, by members of the congregation. They were usually of large size, with from twelve to sixteen branches, and inscribed with name and date. The iron suspending - chains are necessarily very long, and often made a vehicle for rich decoration, taking the form of enormous links fashioned of leaves and scrolls, the design repeating four ways on the same plane. One of the finest remains in Lincoln Cathedral, the principal link measuring about eighteen feet in length. Another, in St. Saviour's Cathedral, Southwark, is held by a chain of decorative smithwork, comprising seven elaborate

* G. H. Birch—*London Churches of the XVIIth and XVIIIth centuries.*

links, increasing in size and importance from the roof downward. In this the design repeats eight times in the important links, four at right-angles to the rest, or four-way,
somewhat confusing the effect. Examples of such chains
are not uncommon, as in St. George's, Yarmouth, but simpler
links, on one plane only, are far more numerous. A remarkable
example, resembling that at Lincoln, is in the Museum (Fig. 43).
In London there appears to remain but one other, in Christ
Church, Newgate Street. There are some in Bristol; one
each in Oxford, Waltham Abbey, Lingfield, Lancaster,
Godalming, Horsham, Walton—in fact, they are still fairly
abundant, though vast quantities have been ruthlessly
swept away by the mediævalist "restorer." In some cases
the churches were lighted by smaller chandeliers suspended
from wrought-iron brackets, as in St. Margaret Pattens.
At St. Paul's, Shadwell, four were suspended from horizontal tie-rods, each marked by a scroll pyramid over the
chain. In rare cases the heavy sounding-boards suspended
over pulpits were held by iron supports of very similar
design to the great links of the chains. The finest, twelve
feet long, was formerly in St. John's Church, Chester, and
another is at Stoke d'Abernon.

English chandeliers of decorative wrought iron are
extremely rare, and there seem to be but one or two extant.
These take a basket form with flowers and wheat-ears,
skilfully and daintily forged, with stems and branches
entwined with flowers and tendrils. The design and execution are excellent, and suggest affinity with the Claydon
stair balustrade, in which the wheat-ears are so poised
that they are said, with some exaggeration, to tremble
when passed by, as if bending to the breeze.

Another acceptable gift to the Church seems to have
been the marble altar-table on richly wrought ironwork,
probably produced by the same smiths, since the forgings

have identical characteristics. For a period iron consoles were in great favour, but they are now rarely seen in churches, though not uncommon in country houses. There is one, however, in St. Andrew's, Holborn, its work and design being based on an extremely light treatment of French rococo. A shell and scroll treatment forms the front with brackets and console supports, claw feet, and a profusion of lightly forged pendent flowers, fruit, and foliage. The sides usually repeat the consoles and scrolls, unless pilasters of vertical bars take their place, as in Communion-rails. There are other examples in St. Olave's, Southwark, and St. Clement Danes. By far the finest, however, are those in All Saints' Church, Derby, and Wollaton Church, Notts, and another at Loughborough. There is reason to regard these as of Derby *provenance*, and the name of Humphrey Yates, of Derby, was given as the maker by one who professed to know. They are certainly more common in Derbyshire manor houses than elsewhere.

Font-cranes are rarely seen in our churches, but three or four earlier examples have previously been mentioned. An example in Allhallows, Barking, is singularly massive and ungainly, and a lighter crane in St. Peter's, Cornhill, is now dedicated to other uses. A fine specimen from the church of St. Michael Queenhithe, is in the Museum (No. 144-1889).

Signs to denote crafts, trades, and houses of entertainment are of the remotest antiquity, the latter in England being at first distinguished by coloured lattices, bushes, garlands, or representations of bunches of grapes. Sign-brackets were favourite objects with French designers, and we have a series of designs by Du Cerceau in 1570-1625, by Tijou, 1693, and Lamour. Of actual specimens of early date there are few. One only, of sixteenth-century date, was, and may yet be, at Bruges. Our earliest appears at Wendover,

and from the long spiral twists laid in the leaves and its
bunch of tulips, it may date from Charles II's time. Notwithstanding Tijou's rich and elaborate designs, or because
they were impossible to execute, few decorative iron frames
or sign-brackets seem to have been produced till well after
the first quarter of the eighteenth century. Dutch and
Flemish painters rarely represent other than plain wood
posts and brackets, from which hang the painted boards,
and few others are seen in English paintings until 1747,
when Hogarth painted in minute detail a sign, supported by
an elaborate wrought iron bracket and stays, in his picture
of the Lord Mayor's Show. Two smiths' signs of wrought
iron exist in the Nottingham Museum, without the brackets,
one with three horse-shoes and the date 1735, and another
with the date 1752, comprising a horse-shoe and set of smith's
tools. Well-designed brackets of the eighteenth century
are still to be met with in country towns and villages, some
with considerable projection, consequent perhaps on relaxa-
tion of rules and regulations imposed when every business
house had its sign, and the number were a danger to traffic.
Most are purely local and English in design, rarely with
acanthus - leaves or embossing in the manner of Tijou.
Brackets such as that at Gretton, Northants, which finishes
in a dolphin's head, are very unusual; a well-known
example is at Canterbury, and illustrations of others
have appeared in various books and magazines. Another,
at Coalbrookdale, is elaborately scrolled, and a striking
example at Much Wenlock has an oval garlanded with vine
on which a raven perches. More notable examples are at
Thame, and at Bruton and Mere, in Wilts. Large numbers of
signs are fixed on the top of tall posts or hang from brackets
fixed to them. These occur on the old coach roads leading
from London, as at Stratford, Ilford, Eltham, and Chelmsford;
one is at Highgate, on the old North Road. Excellent

examples, too, are at Horsham, Aylesbury, and Woodstock. These are usually better designed and more artistic than those projecting from house fronts, perhaps because more compact. Jores published sixteen designs for signs in his pattern-book in 1759: in " Gothic," " Modern," " Chinese," and other tastes; but it does not appear that many were executed, at least none exist. The taste for wrought iron signs of exuberant richness which prevailed on the Continent, especially in Germany, was never shared by us; but, on the other hand, our paintings on some of them were excellent, and became stepping-stones to artistic fame.

Vanes are somewhat akin to signs, and may be traced back to an antiquity hardly less remote. The cock possibly proclaimed the nature of the building from afar in early days. Its various treatments are much conventionalised, and they were in use for the purpose with us even prior to the tenth century. Banners pierced in copper surmounted the turrets of castles and halls, and were similarly fixed as vanes on church towers in compliment to the local baron, even to the exclusion of the more usual religious symbol. The oldest church vane extant in England is that of Etchingham, a banner of arms in pierced copper of the latter part of the fourteenth century. A royal banner dated 1639, with "C.R." and the Royal Crown pierced, is on Ruscombe Church, in Berks. Another, on Kensington Palace, commemorating William and Mary, consisted of " W.M. " interlaced and pierced between two Rs under a crown. One of the finest treatments is the pierced banner vane of Oxborough Hall, Norfolk, 1661. Another noteworthy example, fringed with fleurs-de-lis, with " I.R. " pierced, in honour of James I., 1616, is at Barlborough Hall, Derbyshire. The vane at Blickling, pierced with the eight-pointed star of the Hobarts, is of similar date, about 1620. The vane on Street church, Somerset, has " W.D.," for Walter Dobell, and the date

K

1636 pierced on it. One of the most important
examples extant is that to Lambeth Palace, 1663,
the flag pierced with the archiepiscopal arms, edged
with fleurs-de-lis and two pennons, pointer, and crowned
with the mitre. That to St. Ethelburga, Bishopsgate, is
pierced with " S.E.," and date 1675, with fleurs-de-lis, crosses,
and a cock over the pointer; letters on waved bars below
show the cardinal points. St. Mildred's, Bread Street, by
Wren, has shield and pointer under a crown, with five horse-
shoes upon a chevron, the Red Hand of Ulster, and
interlaced " P.P.M." Other notable vanes were the models
of ships, like that at Portsmouth, six feet ten inches
long, put up by Prince George of Denmark in 1710, and the
model of Sir Cloudesley Shovell's frigate at Rochester about
1708. Several City churches had models of vessels for vanes.
Among City vanes, the grasshopper of the Greshams, eight
feet eight inches long, has always been conspicuous over
the Royal Exchange; also the gridiron of St. Lawrence Jewry,
the comet of Horsleydown, the key of St. Peter's, Cornhill,
the doves of St. Mary Abchurch and St. Andrew Undershaft,
the dragons of Bow and the Guildhall, the swan of St. Paul's,
Covent Garden. Others are surmounted by crosses, or ball
and baluster finials. A remarkable vane is that to Minster
Church, with the grim horse's head commemorating Sir
Robert de Shurland's feat. A few vanes represent fish.

CHAP. X.—EIGHTEENTH-CENTURY ARCHITECTS AND THEIR USE OF IRONWORK.

WREN and his immediate followers left few working designs for ironwork, and none were published by architects until Gibbs brought out a book on architectural design in 1728, at the outset of his career. This comprised several plates of ironwork, including four similar designs for closures to forecourts, which afford a choice without much difference. They were intended to guide and influence the smith, but it does not appear that any of them were actually carried out. Like the work for the Ratcliffe Library at Oxford, published in 1747, they lack interest. Most of the work actually carried out for him, as at St. Martin's-in-the-Fields, was very similar to that of his contemporaries, and apparently by the same smiths. The effect of the publication, if any, was to damp the ardour of the smith.

So far no smith in England had even remotely followed Tijou's example by publishing any of his designs or any kind of advertisement. Names of the remarkably talented and artistic men who followed more or less in his footsteps, and produced the magnificent ironwork of the first thirty years of the eighteenth century, have been recovered from old household or building accounts or by some other chance. Their works as such were never illustrated, and rarely noticed in contemporary writings.

London craftsmen, however, had the inestimable advantage that work original and successful, from Tijou's onward, could for the most part be freely inspected, admired, or criticised, by every one visiting or resident in the Metropolis.

K2

This fact made London, as it had Paris, the centre where the best of everything artistic was produced. Patrons could point out that which pleased them to the smith they intended to employ, or commission the man whose work they most admired. Competitors could copy, vary, improve, adapt, or combine whatever was open for them to see, and thus suit their clients' requirements. This kept the art of the smith a living one for half a century. New ideas, features, and *tours-de-force* were given birth, and added constantly to the common stock. The best craftsmen were comparatively free from dictation or restrictions in those halcyon days, and emulation and rivalry maintained progress at concert pitch. It seems clear that Wren left details to the craftsmen in whom he had confidence, providing their work contributed to the architectural effects he aimed at. Where the smith's work formed no part of his architectural scheme, he had *carte-blanche*. In the matter of stair-balustrades, altar-rails, and generally, architects of those days seem to have selected designs as wall-papers are selected, only varying them with discretion; and thus similar designs in ironwork recur not infrequently in buildings by different architects. A somewhat prolific compiler, Batty Langley, ventured on the publication of illustrations of ironwork in 1736. His "curious designs for iron gates of which we had none so noble yet executed in England" were lifted bodily from a German publication. Though smithing remained uninfluenced, the compilation perhaps paid as a whole, for in 1739 eight more pages of designs of ironwork "of the most exquisite taste" were reproduced from Tijou without acknowledgment, and other plates comprising designs for balconies, taken from Le Clere, "for use where the slightest balustrades would be too massive." These books were probably compiled to sell, and were of little practical value.

About 1750 the *Builders' Magazine* appeared, by · T. Carter, architect; it included some designs for ironwork, and was carried on for several years. François Babin's book of designs found its way to England at about the same time, a few balconies in Mayfair having been suggested by them, as well as several in Spring Gardens, now destroyed. Isaac Ware's *Body of Architecture* appeared in 1756, with designs for railings, not very original, and including Vardy's still existing iron gates to the Horse Guards, made in 1753. Ware, however, in the text, recommends wooden gates for. screens and advocates cast iron for balustrades as cheaper though admittedly inferior to wrought, his own designs for these being more adapted for wood or stone.

Upholsterers seem first to have taken up ironwork in 1760, when Ince and Mayhew published a balcony design, some stair balusters, and brackets; and soon after, in 1763, a " Society of Upholsterers," of the " Golden Buck," Fleet Street, produced " curious new designs in genteel taste," some of which were for ironwork. J. Jores, possibly a smith, published in 1759 a book with twelve plates appropriated from Huquier's designs, and eight others, inferior in drawing, which were possibly his own. A still smaller book was brought out by W. and J. Welldon, smiths of London, entitled *The Smith's Right Hand*, with plates partly taken from French originals called Modern, with others styled, Italian, Chinese, Venetian, and Gothic, these latter, perhaps, their own designs. Jores' and Welldon's books are now rare. All seem compiled to make money, and hardly pretend to be published in the interests of smithing, as Tijou's book and the fine contemporary works printed in Paris were. Their influence in London on smiths as craftsmen is perhaps only now discernible in such things as a few balconies, the lamp-holders to 45, Berkeley Square, etc. and some civic sword rests. The taste of some architects, or at least that of their more wealthy clients, had probably leaned to the French since the advent of Tijou, or even before, and is, as we have seen, especially notable at Canons, built for the Duke of Chandos, and later in Chesterfield House, Mayfair. The designs for the stair-balustrades in many of the country seats of the nobility, and even for a few mansions in London, no doubt influenced by architects, are distinctly in the French taste of Louis XIV.; and the fine gates and screens in the façade of St. Paul's, Covent Garden, under semi-circles of escallop-shell design, lean to that of his successor. Gates not directly connected with buildings remained perhaps for some time longer in the smiths' hands, as in those to Guy's Hospital, Gray's Inn, and the Temple Gardens, which, however, are chiefly interesting for the dates they bear.

Wren, the greatest of English architects, affected Italian design rather than the more florid magnificence of contemporary work in France. His finely proportioned interiors had not been distinguished by elaborately carved decorations, painting, or metal work until he met with Grinling Gibbons, whom he consistently patronised and spoke well of. Tijou, less fortunate, only obtained a share of his work. They had, no doubt, met at Hampton Court, when Queen Mary was personally superintending the laying out of her gardens, with which Wren was not concerned. He had not so far shown any desire to associate rich ironwork with his buildings; thus there is none at Winchester, Chelsea, or Greenwich, nor in his earlier churches as Bow Church, Cheapside, St. Stephen's, Walbrook, or St. James's, Piccadilly. His ironwork had been sparing and simple. Thus, though very lofty stone piers, surmounted by elaborate trophies of arms, were prepared for the gates to Chelsea Hospital, the latter are of plain vertical bars finishing above in arrow heads, and below in blunt points taken through the lower horizontal. The sole decoration is the lock-rail, two feet wide, filled in with round iron scrolls, with smaller scrolls diverging from them, the ends beaten into plain, wide, deeply cupped leaves with sharply waved points. They were most carefully made about the year 1690, after Wren had seen Tijou's work, and are still serviceable. Wren was giving Tijou the great window frames to make for St. Paul's from this time until 1694, when he trusted him to make balustrades for two staircases of the simplest design, and also with those for the several private staircases at Hampton Court, Tijou's works being on the spot. Neither the designs nor execution of these suggest that either Wren or Tijou cared much about them. Wren, however, continued his patronage to Tijou at St. Paul's to the end, giving him practically a free hand in the richer work for screens and grilles, but with due regard to the suitability of the designs. Also part of the work went to Robinson, a smith of more moderate views. Thus, while Tijou's windows were handsomer and more substantial, Robinson's, of the same size, cost very much less: while the gates which Tijou executed for the south porch cost £320, and were not dear at that, Robinson did what was necessary for both the north and the immense western porches at the probably competition price of £238. Other than the work at Hampton Court and St. Paul's, Wren gave Tijou nothing except possibly altar-rails for one or two City churches.

. Work hardly distinguishable from Tijou's continued to be produced after his final departure in the year 1711. The gates for Bridewell are practically facsimiles of his work, but were not presented by Sir William Withers with the marble pavement until 1714, according to Maitland, or 1715, to Pennant. These suggest the possibility that he left successors, and certainly individuals of his name outlived

his departure. The original gates are now in the Bridewell
Office in ˙Bridge Street, Blackfriars; a reproduction of a
panel is in the Museum (Fig. 44). The high standard
of quality, however, was not long maintained, but the
staircase balustrade made in 1716 for a house in Lincoln's Inn Fields, now in the Museum, is hardly inferior
in execution (Figs. 45, 46). Its first-floor landing panel comprises a framed monogram and mask over drapery, between
large buttress-scrolls of acanthus and eagles' heads. Wren,
though restoring Westminster Abbey, 1698 to 1705, where
gates were needed for the choir, did not employ Tijou. Those
to the aisles, in the pseudo-Gothic taste of the day, are of
excellent work, like those to the apse, but of no particular
style ; but two pairs of low gates in the transepts are rich
with embossed acanthus work, and hung from double panels,
each comprising four finely embossed masks. The much
larger choir gates of somewhat geometric design have also
embossed foliage little inferior to Tijou's. HENRY SPOONER
was Abbey smith from 1702 to 1704, and the post was continued until 1713 to his widow, who may possibly have
employed members of the Tijou family. A rich gate to the
garden entrance of Fenton House, Hampstead, built by
Wren, is designed in much the same spirit and with no less
fine embossing. The monogram of its owner, Joshua Gee, fixes
the date as about 1706 (Fig. 47). The overthrow is stepped
like an Antwerp gable, an innovation which was taken up
by London smiths, and the treatment is skilfully applied
to the pilaster tops. Another equally fine gate, apparently
by the same hand, is reputed to have been taken from
Holland House to North End House, Twickenham, and
thence to Wovington Manor, Sussex (Fig. 48). The balustrade
to the steps of the entrance to Roehampton Court, built by
Archer in 1712, is also in Tijou's manner, and finely worked.
Two gates from Canons seem, though of later date, to be

of the same work, and were probably produced before the owner became Duke of Chandos in 1719. Both are somewhat low in proportion to the height. Canons was demolished and the materials sold in 1747, when one of its gates was erected by Alderman Belcher at the Durdans, near Epsom, still retaining the Chandos motto on a Cloth of Estate, supported by acanthus scrolls with eagles' heads ; Tijou's shell motive with acanthus, used at Hampton Court, forms a base to the overthrow, which comprises waving branches of laurel and the Alderman's crest. The piers are wide and richly worked, surmounted by over-sailing scrolled pyramids with acanthus ; and the verticals of the gates are tasselled in the French manner, almost unique in England, with water leaves welded to them in pairs. A second gate was purchased for Hampstead Parish Church, with fifty-nine feet of railing, for £40, increased by no less than £26 for the carriage (Fig. 49). The overthrow is boldly designed, large scrolls and acanthus connecting the pyramids over the wide pilasters, which, like the gates, comprise handsome panels. Another small but singularly rich relic of the sumptuous ironwork at Canons is over a wicket in the gardens at Wotton, Surrey, comprising acanthus and eagles' heads, and still bearing the Chandos arms over Cloth of Estate and motto. Larger gates closing the mile-long avenue from Edgware, broad enough for three coaches abreast, bore the Chandos arms, but have not yet been traced. At Kensington Gore is a small but beautiful gate between pilasters, of the same workmanship.

Here it may be as well to refer once more to the fact that Tijou brought members of his family with him to England who did not quit the country with him, and whose domiciles in London have since been traced to comparatively recent times. Their independence and rivalry may have contributed to induce Tijou's otherwise unaccountable departure in 1711.

The rich and heavy Chandos balustrades of purely French design, and the marble staircase, were bought by the Earl of Chesterfield for his London house, then building. They are of boldly treated scrolls with garlands, acanthus, masks, tendrils, and monograms within garters, only the ducal coronet needing to be replaced by that of the Earl. It neither suggests Tijou's work nor that of any contemporary smith, but may have been by some foreign smith working for Louis XIV. at Powis House in Ormond Street, who probably made the no less grandiose railings in the forecourt of Chesterfield House, in 1747, which match it. The Earl writes as pleased with his bargain : " My Court, my hall, and my staircase will be really magnificent," and again, " the staircase particularly will form such a scene as is not in England."

Hawksmoor made considerable use of a rich style of ironwork by a smith whose name is as yet unknown. Among his earlier works are the altar-rail and gallery fronts of St. Alphege Church at Greenwich, and the altar-rail and pulpit panel of St. Mary Woolnoth, in Lombard Street, and that to the Chapel of University College, Oxford, all produced between 1716 and 1719. His most conspicuous works, however, are the gates and screen to All Souls' College, Oxford, dating from about 1734, rich and florid in design, but not altogether pleasing. The panels and pilasters repeat those he used for the choir gates at Beverley, 1730, in both cases based on Tijou's gates for the Clarendon Building at Oxford, 1712. Another large gate by the same smith is at the foot of the stairs of the headquarters of the Honourable Artillery Company at Finsbury. All three embody in their pilasters a feature used by Tijou in those of the altar-rail of St. Paul's Cathedral. These are of peculiar lyre design, scrolled above and below, with a centre of four bars converging downward, the inner pair meeting at an acute angle, and thence carried on in a single bar vertically to the base. Wren seems to have

employed the same smith for his balustrade to the Queen's
staircase at Hampton Court, 1731, adapted from a design
in Tijou's book, and also reproducing part of the design
of the St. Alphege rail. The same smith seems to have made
the staircase balustrade for Kensington Palace, a trifle less
florid, in 1730. Archer employed him at St. John's, Westminster, using identical pilasters, but with more sparing
use of acanthus in the altar-rail, part of which is of cast
balusters. The stair balustrade to the pulpit is of lyre
design, and dates probably from about 1728, when the church
was finished. His earlier rail in St. Philip's, Birmingham,
1711 to 1715, is of richer design. This smith also was
employed for the altar-rail and other work in St.
Sepulchre's Church, as well as for those to St. Giles,
and St. George the Martyr, Bloomsbury, all three rails consisting of a four-centred design within a circle, lyre balustrades,
and Tijou-like pilasters, which evidently remained popular.
The dates of these are between 1733 and 1736. The pulpit
stair-rail of St. George's, Hanover Square, is similar. The
altar-rail design reappears at St. Neots, and may have
become a sort of stock pattern. Another rail, saved from
St. Matthew's, Spring Gardens, and now in St. Martin's-in-
the-Fields Church, recalls in its gates that to St. Alphege,
but these are between sturdy pilasters of vertical bars with
acanthus caps of the favoured French design, while the rest
is of lyre baluster panels. The pilasters also recur in the far
richer rail of St. Botolph's Church at Boston, Lincolnshire, in
which the panels are scrolled with acanthus, and the gates are
of radiating scrolls within a circle. One of our most elaborate
altar-rails may be seen in the Grosvenor Chapel, South Audley
Street, possibly made in 1743. Delightful grilles to
windows and doors in the City, now no more, but once
fairly abundant, were of similar designs and dates, but of
stronger make.

Though Vanbrugh used little ironwork at Blenheim, he made considerable use of it at Kimbolton. Here the gates between fixed panels and stone piers, are lofty and severe in treatment, with three fine pyramids above, and centre with the ducal arms. The staircase balustrade is less impressive, but assumes more richness in landing panels, centring in coronet, monogram, and laurel · branches. Richer balustrade work is on the steps to the main entrance from the inner court. A graceful stair rail of husks and flowers and much-beaded scrolls with crinkled water leaves, is at Beningbrough Hall, built by Vanbrugh early in the eighteenth century. Other fine work by him, or by Hawksmoor, is at Easton Neston, Northants, comprising railings, screen, fine balustrades to steps, and staircase ; with monogram, coronet, acanthus leaves and rosettes to the landing panels. These date from about 1713. Gates and screen to the courtyard, and the rich balustrade to the staircase in the great hall at Grimthorpe, date from 1720. For Vanbrugh also were made the rich and graceful stair balustrade and landing panel of Audley. End in 1721, when he modernised the stairs. Unfortunately, " the large and wide iron gates opening into a spacious Court Yard," seen there by Defoe, no longer exist. The lyre balusters, clothed with acanthus, saved when Duncombe Park was burned, were doubtless made by the same smith for Vanbrugh, who was certainly an important and consistent patron.

The Dutch Captain Wynde acted as architect, and was patronised by the Earl of Burlington. The house built in 1686 for Lord Powis, in Lincoln's Inn Fields, and later sold to the Duke of Newcastle, was among his works. The gates to the forecourt are seen in engravings to have been essentially French in design. The house built for John Sheffield, Duke of Buckingham, on the site of the old Mulberry Gardens, had a forecourt " encompassed by an iron palisade," to which the Mall, with its old avenues of limes, formed one of the approaches. The gates are represented in contemporary engravings to have been similar as to proportions with those of Powis House, but presenting peculiarities, notably eight square panels of quatrefoils. The somewhat low pyramidal overthrow displayed the Duke's monogram, garter and coronet. George III., however, replaced these by plain iron railings. Identical setting-out is repeated in the remarkable gates to the forecourt of Lower Lypiatt Hall, Gloucestershire, built in 1705, to which a curiously German touch is given by the wide lock-rail of interlacing scrolls, the numerous whorls of the scrolls comprised in the overthrow, and its height, which rises to a pinnacle comprising the owner's name " Cox," worked into a monogram. This was Judge Cox, traditionally reported to have reprieved his smith, a criminal, until the work was completed.

In these works it appears that architects employed smiths as skilled craftsmen and designers, and entrusted them to provide whatever was required, merely under a general supervision. This becomes first apparent in Tijou's work in St. Paul's, under Wren, which is more restrained and differs considerably in feeling from the designs he published, or his work at Hampton Court. Probably all the works here described were similarly influenced, though certainly not designed in detail by the architects for whose buildings they were executed.

Later in the century the brothers ADAM steadily pushed their way to the front. Robert held the appointment of architect to George III. from 1762 until he entered Parliament, and was the author of many fine publications, and an artist. James was also much patronised by the King, especially at Windsor, and by the nobility. Their taste was refined, and came to be based almost entirely on the Classic and Italian. The Royal patronage was followed

as usual by the Court, and the Adams' influence eventually
seemed to carry all before it. They assumed the functions
not only of architects, but decorators, and if the client
permitted, would not allow so much as a picture or piece
of furniture to be placed without their approval and consent,
or until complete schemes had been drawn out.

A large number of original drawings by Robert Adam
are in the Soane Museum, many being for gates and grilles.
The gates, however, are more noticeable for the beauty and
dignity of the stonework, arches and colonnades, in which
they are set, than for the ironwork, which may be partly
cast, with enrichments both of bronze and lead. Smiths' work
as such did not appeal to them except as a means to an end.
These evidence a complete mastery of detail within the
lines laid down, for even three or four alternative schemes
for the treatment of but a small and inexpensive iron gate
are met with. We find Robert Adam engaged in 1778 in
designing entrances to Hyde Park. In one of these the chief
feature of the gates consists of bars crossing diagonally,
with rosettes at every intersection in the manner of contemporary French *treillage*, interrupted by pilaster-like
verticals of a repeating vesica, with a corresponding
frieze above. This supports a pyramid comprising the
royal arms, garter and crown, with vertical work
under the lock-rail. These were within a fine stone
arch, with three lesser arches and simple gates and
railings on either side. An alternative design for the
Hyde Park entrance recalls the gates at Sion House.
No new gates seem, however, to have been erected till 1828,
when the existing design by Decimus Burton was carried
out. Of other designs by Robert Adam, two dated 1776
have the verticals above and the *treillage* below the lock-rail.
In gates for Osterley, 1777, an arching stretcher crosses the
verticals, sustained by a small arch between each pair ;

while above, the verticals, duplicated by enriched intermediate bars, give an open look to the region below, reversing the usual scheme. Gates fixed in 1782 for the Hon. R. Rigby, of vertical bars above the lock-rail, have a broad lozenge band below, with honeysuckle and geometric filling, opening under a large fixed grille of radiating design with enrichments. These were an improvement on very similar gates made in the same year for Lord Coventry at Croome Court; almost repeated in 1784 for the Rt. Hon. C. Loftus. In the same year large and rather plain gates with arms and coronet, were designed for Lord Dumfries, in which big circles of radiating ornament are the feature of the wickets above the lock-rail, in the manner of sixty years earlier. Gates, in which the lower parts were of trellis design, were drawn for Lord Wemyss at Gosford, under a triumphal arch, surmounted by a monogram and coronet, an almost unique instance. Most of the designs are for gates within imposing stone arches and colonnades, while those to stand between piers are usually of vertical bars with semicircular stretchers. Enrichments of the horizontals take the form of bands of circles, half-circles, or lozenges. A design, of which there are several examples, ends the verticals abruptly on the stretcher, the region below being occupied by one of the Adam fan-like designs. Robert Adam was a prolific and facile designer, with a complete mastery of details suited to his purpose; it is impossible to convey more than a general idea of his range, though his designs are easily recognised. The ornament is delicate, refined, and exclusively based on the classic; somewhat dilettante, not to say effeminate. Their architecture and decoration held the field while the brothers lived, and would-be competitors and producers followed in their footsteps so closely that wrong attributions may be possible. Conformity to their standard of design became the paramount necessity, so much so that the originality which distinguished the earlier craftsmen became suppressed or dormant.

It would appear that relatively few of the Adam designs in the Soane collection were actually carried out. Those that were so would pass into the clients' possession, possibly to be preserved, as at Kedleston, or as the Drapers' Hall designs in the Gardner collection, and many appear to be preliminary or trial sketches. Perhaps the most important of those executed and remaining *in situ* are the gates of Sion House, Isleworth. This design is chiefly of plain vertical bars, with lighter bars dressed with water leaves and opening into lozenge spaces between, filled in the lower part of the gates with cast rosettes; the tops are elliptically curved, falling towards the centre, with spikes and tulips alternating over scrolls, without central feature or overthrow. The rosettes are of cast bronze and iron, the remainder wrought iron. The gates are set in a noble stone arch now surmounted by the bronze lion removed from Northumberland House, with two graceful lamps. A stone colonnade of six columns on either side is closed by grilles and surmounted by lamps of similar design. A London example is at Lansdowne House, the garden entrance consisting of a small pair of gates with horizontal top, and verticals with cast javelin spikes, the upper and lower horizontals and lock-rail duplicated to receive circles or ellipses. Below the lock-rail are ellipses of fan-tracery centring in lions' masks, while the four-sided piers are filled with a honeysuckle design capped with beehives, the Lansdowne crest. These gates are of relatively early date, 1765, made to the order of the then Lord Shelburne. All the work is cast, unless where cheaper to forge.

Some of the most decorative of the London area rails are from designs by Robert Adam. A house built on the east side of St. James's Square in 1773, for Sir W. W. Wynn, still preserves its handsome railings. The verticals rise above the horizontals in mace-heads over a honeysuckle border, and the base is formed of fleurs-de-lis within circles. The pilasters are of lyre design, with fan and acanthus enrichments. The finished drawing for these, in the Soane Museum, shows all four pilasters surmounted by rich standards with lamps. Winchester House, next door, has a railing of javelins with borders of ellipses, and retains its dignified scrolled obelisk lamp supports. A house on the north side of Portman Square, built by R. Adam in 1775 for Lady Horne, also preserves the original railing, fortunately with its lamp standards complete. The verticals rise in moulded spikes through double horizontals, each space with three compressed ovals repeated near the base with arrow points above and semi-circles below. Between each pair of bars is a central lozenge and circle with moulded spikes. The pilasters are of open-work baluster design, enriched with honeysuckle, etc., and the outline of the lamp standard forms an obelisk with fan, honeysuckle, and looped ornament. The railing to Chandos House, in Queen Anne Street, also with the original lamp standards, is enriched with four borders of arrow-points with leafage and loops, and one of rosettes below the moulded spikes. The pilasters are of handsome honeysuckle design, with lyre pattern lamp pedestals to correspond. A different Greek honeysuckle is used with excellent effect in the pair of lamp-holders in John Street, Adelphi, each comprising three repeats of the honeysuckles on a central stem, entirely of wrought iron, and thus probably unique. The fine mansion at the corner of Harley Street formed the west wing of the Duke of Chandos's intended dwelling, and became the residence of the Princess Amelia. It has a relatively plain railing with javelin-heads and diagonal border, and supports several lyre-shaped lamp pedestals with delicate lozenge and honeysuckle enrichments,

L

a turned baluster support taking the place of the usual enriched pilaster below. The simple railings to Boodle's Club, in
St. James's Street, are a not unpleasing contrast to these elaborated examples. No doubt many of the severely plain railings
of stout vertical bars, like those on the south and east sides of
Fitzroy Square, are by the Brothers Adam, especially such
as include cast vases to the standards, suggesting Adam designs. In their fine screen to the Admiralty they appear to
have allowed the older railings to be reinstated. Perhaps
the most dainty balustrade of the period in London is the
richly worked rail to Guy's tomb in his hospital, of lyre
panels, with rhythmic and graceful arrangement of leaves
and husks in Adam's classic manner, between a low base of
closely set verticals and a border of circles above; these are
connected by leafy festoons, a honeysuckle between each pair.

The Adam balustrades were always rich and refined.
A drawing in the Museum represents one formed of a Greek
honeysuckle between deep Greek fret borders, with pilasters
to match, bearing lamps apparently designed for a bridge.
This terminates against a carved stone pedestal and sphinx.
The balconies at Boodle's Club, St. James's Street, afford
early specimens of Adam design, about 1765, and consist
of light and closely set bars embellished with alternate
leaves and swellings, every fourth bar bearing a cast oval
patera near the base. Trellis designs for balconies were
perhaps introduced by the Adams, but they made no great
use of them in their London houses. On the other hand,
they excelled in their designs for grilles over doors needing
protection. Two superb specimens have recently been
rescued from buildings pulled down, and placed in the
Museum, both finely designed and of excellent workmanship.
This is well seen in the beautiful grille from Drapers' Hall,
now lent to the Museum, accompanied by the original
drawing, with the memorandum that it is of wrought iron

and brass, and the approximate price £13 13s., which adds
to its interest (Fig. 50), and in that from Harewood House, no
less remarkable. No. 13, Mansfield Street, built about 1770,
preserves its door grille, the radiating design of ironwork being
intersected by an arch of wood and plaster. Two other
fine examples on the north side of Portman Square, and several
in Portland Place, built by the Adams in 1778, would be
designed by them. Many original designs for door grilles
signed by Robert Adam are in the Soane Museum. Of
their stair balustrades there are still many in London, but
perhaps more remain in country mansions, like that of Gos-
forth House, the seat of Lord Wemyss. Typical examples
are represented in the Soane drawings. Most, though slight
and delicate in execution, have proved durable. Excellent examples are those to pulpits in churches, as at St.
Bride's, and there are many altar and chancel rails, some
showing incipient concessions to Gothic feeling. On the
other hand, large quantities have been cleared out in so-
called " restorations," including several formerly in Westminster Abbey. Relatively few lamps and lamp-holders
exist, for they were for the most part slight but graceful, and could not readily be adapted for gas. This
accounts for their destruction in the Adelphi, built by
the Adams as a speculation. Excellent designs for lamps
combined with balustrades are in the Soane Museum, notably
one for Cumberland House, Pall Mall. One is especially
bold, and comprises two winged terminal Cupids holding
torches on either side of scrolled honeysuckle supports;
and four or five of the others are also remarkable. Perhaps but one pair still exists in its original position, on
either side of the front door to Sion House, the design
being enriched with honeysuckle, Greek wave, rosettes, etc.
Dublin is richer in work of the Adam period and character;
and much may probably be hidden away in Edinburgh.

L2

CHAP. XI.—CAST IRON WORK.

THE fact that, given sufficient heat, iron melts like every other metal, must have been known in mediæval days ; but its quality in that condition being worthless either for armour, weapons, tools, fastenings, hinges, grilles, or any purpose then needed, it had no commercial value. Jhone Colins, an ironmaster of the Weald, was the first to conceive the idea of covering his grave in Burwash Church with a large slab of cast iron, on which he placed the inscription in Lombardic letters, " Orate p annema Jhone Colins," without date, but over the name is a small fourteenth-century cross. The letters are those current in that century, but the actual date of the casting can never be certainly known.

It was not until the reign of Henry VII., when the old louvres with outlets for smoke in the roof, now so rare, were being replaced by hearths and chimneys, that any extensive use for cast iron was found. A demand arose for cob irons, or, as they were more politely termed, " andirons," derived from the French *landier* or *andier*, which in France were then richly designed and cast in iron. The French model was tall and somewhat tapering, of pilasters terminating above in some human head under a moulded cap rising from a low arched base ; the entire front richly worked with conventional tracery, foliage, etc., generally embracing shields of arms or the sacred monogram. These designs were not slavishly followed here, but our nobles required andirons no less rich in character, and thus a demand was created. A fine pair at Castle Hedingham end above in female heads in the coiffure of Henry VII.'s reign, and are decorated with tracery; and a simpler pair is at Northiam, both without the French taper, which is, however, seen in other English early examples. At Helmingham, in Suffolk, is a pair with angular stepped bases each surmounted by a lady with somewhat different head-dress, bearing shields. The old English term "Brand-iron" was revived for these in the Weald. At the same, or not much later period, cast iron fire-backs came into request, chiefly at first for royal use. The decorations for these are modelled or carved on small rectangular blocks, and the earliest consists of the royal arms, a shield with "E.H." over a fleur-de-lis, and a four-leaved rose; all three beneath the very low crown used for a time by Henry VII. The "E.H.," it may be remarked, fixes the date as 1487, when Elizabeth of York was first crowned with great pomp as Queen Consort. Besides these were a larger crowned lion *passant regardant*, and an uncrowned lion *passant*, to face right and left; while some meaning may attach to two small nude grotesques impressed from a separate block. From these the royal designs were constructed. They present the two crowned shields, one over the other, in the place of honour, flanked by the *passant* lions, with anything handy as required, blank spaces being unappreciated. A rope edging was pressed round the sand mould completing it for casting. Other accessories, as a long dagger and a four-leaved rosette, were used, and short pieces of rope, plain or crossed, came in handy to fill in blank spaces. Several variations of this back are known, adapted to different sizes (Fig. 51). A slightly later lavish scattering of fleurs-de-lis uncrowned, singly or in groups, may relate to Henry's claims and war on France, when, being bought off, he returned triumphantly with full coffers in 1492. A rarer example shows the Tudor rose issuing from the pomegranate, evidently alluding to Prince Arthur's marriage to Catherine of Aragon in 1501; this is used twice, and the rest *semé* with rectangles bearing the dragon of Wales. Fire-backs for sale were similarly arranged with meaningless decoration without royal attributes. One at Waldron, over five feet wide, required seven crowned *roses soleils*, four rams, four rosettes, and various crosses, concocted from rope-ends ; and another has ten fleurs - de - lis and rope - ends ; and from these we descend through ruder and simpler forms to the absolutely plain. It seems that the local gentry and farmers might choose their stamps and have the designs composed under their eyes, to any size required. Several are impressed with andirons, one bearing the initials of Henry Neville, ironmaster of Mayfield. The richest fire-back of this date in the Museum is entirely covered vertically with strips of Gothic vine pattern wood-carving, and the badge or rebus of the Fowles, also ironmasters of the Weald (Fig. 52). Wood-carvings, often mere scraps, were frequently pressed into the service. A grave-slab in Rotherfield Church of this date, without inscription, bears two crosses.

That iron-casting as an industry originated in the Weald cannot be doubted, and it must thus be conceded to have originated in England. It did not, however, become really important and flourishing in the modern sense until 1543, when Ralph Hogge, or Hugget, had the happy idea of interviewing Henry VIII. As a Sussex ironmaster he offered to cast cannon in iron for him, which hitherto had only been cast in bronze. His price was 10s. per cwt., while bronze was costing 70s. Henry, with his mind full of French and Scottish wars, was just then requiring guns in immense quantities, and even his enormous revenues were feeling the strain. Thus Hogge's offer was welcomed, and Henry's foreign experts, who were casting bronze cannon in the Tower, were sent to Sussex. Hollinshed records that cast iron cannon were produced this very year, but Lord Herbert of Cherbury gives the following year, 1544, as the date when iron pieces and grenades were first cast in England. A little later the proportion of guns in the Tower was 351 iron to 64 bronze, and the *Harry Grace de Dieu* carried 102 guns of iron to 19 bronze. The fortunes of the Weald were made, for an immense export of cannon and shot ensued, and many of the great nobles of the land embarked in the business. It became almost a Court monopoly, and only ended when the woods and forests of the Weald were entirely destroyed.

Besides cannon, Henry supplied the founders for the first time with specially carved wood patterns for his royal fire-backs, on which his arms and supporters were properly and heraldically displayed. One of these, among the earliest examples, is a back with moulded edge, and dragon and greyhound supporters within a three-centred arch. Andirons of massive architectural character were also cast for him, of which examples are at Leeds Castle, Kent, bearing rosette, crown, chalice, and fleur-de-lis on the front ; and another no less massive in design is at Groombridge.

Edward VI.'s reign, begun in 1547, brings new designs for fire-backs, on which his royal arms and garter are displayed under a crown, departing considerably from heraldic orthodoxy (Fig. 53). On another is a roundel with the royal arms within a garter under a six-arched crown. There seem to be two or three versions of this. Otherwise the most usual fire-back is that of Henry VIII., but with the addition of side pieces and the initials " E.R." The best known back of this reign is, however, the salamander issuing from flames, the arms of François I., dated 1550. In the following year Edward prohibited casting in the Weald, in consequence, according to Wriothley, of a dearth of wood and coal, but nevertheless a back with his arms within a garter appears with the date 1553.

Under Queen Mary the backs produced are few, the most

notable representing a man and his wife being burnt at the
stake, about 1558. None apparently bear royal arms, except
one with the arms of Spain. Others present two heads issuing from a flaming kiln with an angel above : and St. Paul
holding a viper, with a fire burning below. These seem
inspired by the terrible religious dissensions in this un-
happy reign.

Elizabeth reverted to the older Tudor supporters, lion and
greyhound, for her royal arms, on a rectangular moulded
slab arched centrally to receive the crown, which is depressed.
At least three other ruder versions of these arms exist,
with " E.R." An apparently private back bears a large
Tudor rose on a plain shield with motto, garter, supporters
and crown, the edge moulded and top arched in a semi-circle,
with date 1571, and initials " T.K." More beautiful and
rare is the panel with large monogram of enriched " E.R.,"
gracefully roped together, under an arched top, with date
1565 ; buttressed by decorative scrolls from which proceed
tulips, over a plain base. Of fire-dogs a fine pair is on scrolled
and moulded base, with " E.R.," supporting a caryatid
with Ionic cap. Far superior in many respects are the fire-
backs specially modelled for the nobility, which display
elaborate and diverse coats-of-arms and supporters. Foremost are the Cowdray backs of the first Viscount Montague,
with sixteen quarterings, supporters, helmets, mantling, crest
and motto. Two models exist, the more beautiful under a
moulded semi-circle entirely filled by the helmet rising
from a viscount's coronet, with winged crest and a
profusion of ostrich feathers. Of these, seven more or less
perfect examples are in public collections. Other families
are also commemorated by fire-backs of large size, with
small private shields repeated to cover the whole surface,
sometimes with ten or more repeats, and these were presumably sold to the public, and are far from rare.

The Nevills had backs cast of different sizes, with their talbots within a wreath. Occasionally a fire-back records a marriage or death, by name and date in full, or by date and initials. Biblical subjects are much in evidence, crowded with figures and events. The andirons are also interesting and well designed. A fine one, of which there are numerous examples, supports a shield of arms and is dated 1583, commemorating the family of Cromer. Another bears the Ashburnham shield and same date. The most modern as to design is an obelisk on scrolled feet, ending above in an Ionic capital with " E.R.," and date 1573. A rare and perhaps earlier dog is at Charlton House, Kent, of three ellipses piled between neckings, the surface tooled in a honeycomb pattern and partly twined with ribbons, on scrolled feet and shield. Several interesting examples in the Museum are illustrated in Fig. 55.

In Crowhurst Church is a richly decorated grave-slab with arms, a purse, and two kneeling boys and two girls, with an inscription recording the death of a daughter. The same is commemorated in a fire-back, of which there are many replicas, of the year 1591. A cast gravestone with shields only is in the Maidstone Museum.

The cannon of Elizabeth's time are finely cast, clean, accurate, and long in proportion to the bore, and the Weald became busy when our shores were threatened by the great Armada of Spain. A notable example, dated 1590, is at Pevensey Castle, with " E.R.," and the Tudor rose and crown. The industry was then, perhaps, in its zenith and becoming almost a courtier's monopoly. Guns and shot, of which there were about fourteen different calibres, were clandestinely shipped to enemy ports.

With the accession of the Stuarts no change in style was inaugurated. A back with royal arms and Tudor supporters has only the " E.R.," and the date changed, to suit the incoming of James I.; but in the same year, 1604, a new one was ready, with the lion and unicorn. Several variations of the royal arms were produced in competition by different founders, but only one is finely designed, with a semicircular top and high imperial-looking crown of eight arched supports for the orb, dated 1621. Soon after, a number of dogs and backs are found with the fleur-de-lis, probably in token of rejoicing that the Spanish match was off, and the engagement of Henrietta Maria of France to the heir to the throne, announced in 1623. Most of these were decorated with three fleurs-de-lis, but one at Hampton Court is powdered with upwards of one hundred. Another bears a lion crowned, below a thistle, and on either side the crowned rose of England and crowned fleur-de-lis of France. A Prince of Wales back also appears for the first time. The backs for nobles are as fine as in the preceding reign, notably that of the Dacres of Hurstmonceux, with its twenty-two quarterings, or that for the Francis family, 1606. One of 1622 bears the Percy crescent and coronet, with "H.N.," and branches of acorns and pomegranates were added when the Spanish match was favoured.

Several slabs of different sizes are covered with the oft-repeated shield of the De la Warrs. Two are dated 1627 and 1629, both with initials apparently added for the purchaser. The Pelham buckles, with and without the strap, are similarly used. Decorations compiled from separate blocks in the old way remained in demand, such as the bayonet and fleur-de-lis, but especially initials and dates. Other designs bear stags, well carved, or vases with flowers and scrolls. The andirons maintain the standard, and are similar to those of the preceding reign; and one may be specially noticed bearing a bearded figure in relief, with a large clay pipe and tankard, presumably of ale, over a shield with phœnix. The sacred monogram is also somewhat prevalent on andirons, but the taste for backs with religious scenes had almost died out. Of grave-slabs there are five in Wadhurst Church: two, 1612 and 1624, with simple inscriptions; one with a shield of the Barham arms impressed six times, and inscribed "W.B., 1617"; another with three shields and " I.D., 1625 "; and the fifth with six shields of the Fowles. A slab is in West Hoathley Church, 1635, one in Cowden Church, Kent, 1622, and a very remarkable slab, 1638, in the Folkestone Museum.

No back of Charles I. with the Royal Arms of England bears an earlier date than 1635, with " C.R.," but undated they are not rare. As in the previous reign, the first and fourth quarters bear the lions and fleurs-de-lis quartered, the arms of Scotland are in the second, and uncrowned harp of Ireland in the third. A shield with three fleurs-de-lis on scrolled background under the royal crown, for the Queen Consort, seems to have been very popular. Sometimes these have stags addorsed for supporters, stags having at one time supported the arms of France. Among arms of the nobility is one with a large pheon and coronet, date 1630, more usually altered to 1647, for the then Earl of Leicester ; but the finest of the many of this reign is that of the Villiers, Dukes of Buckingham, with helmet, garter and mantling. Another interesting back depicts Richard Leonard, of Breede Furnace, hammer in hand, and his dog. A view of the furnace and some of the tools .in use, and on a shelf three finished vases, form part of the background, 1636. Fire-dogs now become rather uninteresting, a half-length figure, nude, with waved hair, over a pedestal with fleur-de-lis, shield of arms and date 1640, being among the most common. This was cast in Burwash and subsequently brought up to date 1690, by obliterating the shield.

Commonwealth fire-backs are not numerous. The best-known has a clock dial, 1652 ; a design followed more scrupulously by some rival a trifle later. The only political or personal allusions are a General Fairfax galloping triumphantly; and a Perseus riding to attack a monstrous head, with the words "*Nil desperandum,*" and an uncrowned lion, 1656. Three City Companies have fire-backs with their arms—Vintners, Girdlers, and Blacksmiths. That for the Vintners, with their shield environed by grape vine, is modelled by Lesueur, a French artist in some vogue. Of backs alluding to religious disputes, a divine removing the Pope's tiara is a poor specimen. Of family backs, one with the Maynard arms deserves notice; its date is 1635. A back with long churchwarden tobacco pipes and glasses represents social life. Four grave-slabs to the Barham family, already mentioned, are of Cromwell's date. Two models of backs with anchors, and the touching inscription, "PROBASTI ME," 1656, certainly look like personal appeals, and one slightly later, with an anchor and some sort of crown, seems to follow these up. The Restoration brings in a crowd of backs expressing the general joy. Some with simply the Stuart royal arms without initials or date may be of this time; but there is no doubt as to the back with two large C's in monogram under a royal crown, with palm leaves and garlands of laurel, 1661, commemorating Charles's accession. Another of the Royal Oak, with three imperial-looking eight-arched royal crowns and "C.R.," may keep in remembrance past dangers and adventures (Fig. 54). He is also represented bare-headed on a galloping horse with "C.R." A more ambitious back, also with Charles on horseback, trampling on a nude helmeted figure, on a pedestal, with the date 1672, places him beneath a canopy upheld by twisted columns; above are two winged female figures and the royal crown. The figure has been translated as "discord" and "intolerance," but 1672 is the year in which Charles joined Louis XIV. in declaring war on Holland. Another

back represents Charles as St. George slaying the Dragon,
also between columns, with cherubs above holding palms.
A back on which squares with the crosses of St. George
and St. Andrew intersect, impressed four times, may fore-
shadow the Union Jack, which Cromwell had endeavoured
to bring into being, fleurs - de - lis between indicating
1672. The same crosses intersect a square on a shield
with fleurs-de-lis encircled by a collar under an imperial-
looking crown, with palm branches, 1681. One dated 1679,
with " *Pax* " upon it, may mark the passing of the Habeas
Corpus Act, a bulwark of English freedom. A prophetic
Phœnix back, 1664, antedates the Great Fire by two
years. Andirons were rapidly falling into disuse, but a few
with caryatids or masks may be of this reign. Of
Charles II. grave-slabs there are thirteen in Maresfield
Church, one only having shields of arms, to a Barham.
There is a rude one in Sedlescombe Church, and a
cast iron slab in St. George's, Botolph Lane, recording a Lord Mayor. A rare Freemason's back, the very
last of those made from stamps separately pressed in the
mould, deserves passing notice. During this reign the much
more lightly cast and richly decorated fire-backs from Holland
became imported in quantities, but patterns of these were
bought up by founders of the Weald and soon formed their
stock-in-trade. The backs of James II. are entirely of this
class, and the two latest cast iron grave-slabs, dated 1688,
are of his reign. All these are higher than wide, arched above,
usually of a centre panel with a symbolic, mythic, biblical
or heroic subject having no reference to current events,
or very often simply of groups, vases, or baskets of flowers,
of which tulips are the most favoured. Surrounding this is
a wide border with mouldings of flowers, fruit, or scrolls.
Over all is usually a sort of rich canopy of floriated scrolls,
drapery, dolphins, etc. Dates are rare, mostly of the third

quarter of the seventeenth century. Occasionally a monogram may disclose initials of a designer or founder, but these, like the dates, were soon abandoned as hindrances to an unfettered sale.

With James II., casting iron in the Weald on a large scale came to an end. The growing scarcity of timber, which had been felled in the most extravagant manner, had driven the founders to emigrate, and a gradual movement to South Wales had for long been in progress, many of the patterns being also transported thither. But when grates superseded open hearths with dogs, the large separate decorative backs became obsolete, and this particular industry died out under the Hanoverians. The iron-workers of the Weald who remained seem to have eked out an existence by making such things as tobacco tongs and other trifles, which they disposed of to neighbours and wayfarers, down to about a century since.

APPENDIX.

CENTRES OF IRON MANUFACTURE.

OF smiths' wares, some of the earliest, apart from weapons, were the horseshoes, with which the noble family of Ferrers is indissolubly linked. Henry de Ferrers came over with William the Norman as commanding the farriers. The Ferrers were previously well known as *barons fossiers*, or mine-owners, in France, and derived the name of "Ferriers" from ancient and valuable ironworks owned by them in 1035. This Henry received Northampton as a fief, and became Lord of Oakham and its castle, and by privilege, granted by Edward I., the family claimed the fine of a horseshoe from every nobleman's horse coming that way. Evelyn, among others, notices the several gilded and curious horseshoes nailed to the castle gate, including one from the palfrey of Elizabeth. The deeply-rooted superstition that this brings luck dates further back, at least to Roman times, and extends to India. Both St. Galmier and St. Lo were farriers, but more turbulent leaders and rebels than saints are found in their ranks.

Sheffield may have been a seat of ironworking even in Roman times, and remained so ever after. In 1379 the incidence of a poll tax fell upon no less than forty-eight hearths in Sheffield, and families named de Smethe, or del Smythe, also such names as John Locksmythe, John Tripot and Hugo Farro, tell their own tale. Chaucer's " Reves tale," with " A Shefield thwitel bare he in his hose," carries back to the fourteenth century. In the time of Henry VIII. the extensive district of Hallamshire, extending several miles west of Sheffield, is said by Leland to have been no less celebrated for smiths and cutlers than Sheffield itself ; Rotherham also held very good smiths for all cutting tools. In the famous ballad of the dragon of Wantley, of Elizabeth's time, the hero bespeaks new armour with spikes in Sheffield town.

In the Henry VIII. Custom House Book, knives were imported from Almayne, Collayne, and France. Elizabeth in her fifth year imposed restrictions on these imports, but, to favour Sheffield, rules were established two years later when, according to Camden, six hundred cutlers were incorporated as "the Cutlers of Hallamshire." Throughout the sixteenth and seventeenth centuries references are found in household accounts to purchases of cutlery from Sheffield. The Earl of Shrewsbury presented Lord Burleigh with a case of " Hallamshire whittels, being such fruits as his poor county afforded with fame throughout the realm ; " but, according to Fuller, in the time of James I. Sheffield cutlery had ceased to be the best, and knives were sold as low as 1d. each, " for common people's use." In 1624 a Bill passed Parliament to put trade on a proper basis, when the existing Cutlers' Company was established, their Hall being built in 1638. In 1633 John Rawson became famous for selling cases of two dozen knives for 30s. By 1670 the number of Sheffield hearths had risen to 296, when it became the staple town for knives. Early in the eighteenth century some 6000 hands were employed, and 1500 tons of Hallamshire manufactures went yearly from Sheffield to Doncaster, and thence by water to Hull. The export trade, however, remained in the hands of London merchants, who had their factors in Sheffield, where the larger concerns were the forges for preparing the raw material. The prosperity of the town was undoubtedly helped by the propinquity of Rotherham, which put up works for forging, slitting, and preparing iron and steel for the Sheffield cutlers. It was one of the first to produce steel and tinned iron plates, which were worked up in the town. This iron was imported, our ores, then found at a depth of fifty to sixty feet, not being adapted to the purpose. Coal was then procurable at a depth of 120 yards. In 1757 Sheffield opposed the importation of iron from America. According to Beckman, steel was first cast there at Attercliffe, in 1770.

Ripon was noted for spurs, still one of the badges of the town, and these and stirrups were often veritable works of art. Down to Henry VIII. some lands were still held at a rent of gilt spurs, and knights were ceremoniously invested with them, and degraded by having them hacked off. "As true steel as Ripon Rowel," was a saying in the days of Elizabeth. Defoe also credits it with still furnishing the best spurs and stirrups.

For a time Gloucester enjoyed a large trade in knives, scythes, shears and cutting tools, when Sir Basil Brooke was the great steel-maker of the county. He held a patent under the Tudors prohibiting foreign steel imports, only revoked when it was found impossible to make razors, lancets, and the finest kinds of knives from English steel, that of either Damascus, Spain, or Flanders being found essential for such purposes. Glaives were made in Wales as far back as Richard III., and bucklers at Wrexham under Henry VIII., while Ross, in Herefordshire, was famous for smiths in Camden's time.

From the twelfth to the sixteenth century, Nottingham had a staple industry in bolts, bars, hinges, fastenings for windows and doors, tools, nails, harness, etc. The still existing names, Bridlesmith and Girdlers' gates, and Smithy Row, indicate where some of the trades were carried on. Dering speaks of the city as famous for its smiths and their skill; hence Fuller's "The little smith of Nottingham, Who does the work that no man can." All such work had ceased by the time of James I. through the introduction, in 1589, of stocking-frames, each needing 2,000 pieces, mostly of iron, giving occupation to all the smiths, even while discouraged by Elizabeth. Camden notes that the smiths were using coal, though Sherwood Forest still supplied good stores of wood. When, in 1669, 700 frames were at work, Charles II. granted a charter to a Guild, but even before the introduction of steam, in 1838, these had increased to 14,000.

Belper and Dudley became famous for nails, the latter supplying those for the building of Hampton Court. An item in the Record Office runs, "Payde to Raynalde Warde of Dudley, for 7,250 of dubbyll tenpenny nayles inglys at 11s. the 1000." The iron used in all these places was from South Staffordshire, which Camden cryptically remarks had much pit coal and

M

iron, " but whether to their loss or advantage must be left
to the inhabitants themselves as they are the best judges."
Stourbridge, now so celebrated, did not come into note for iron
and glass works until the time of William and Mary.

Birmingham is in Domesday Book, and took its name from
the noble family who owned it. Little is known except that
tool-making was established there before the fifteenth century, when the iron came from Stafford and Warwickshire,
and the coal from Wednesbury. Leland approached it through
a hamlet called Dyrtey, where smiths and cutlers dwelt, and
found many smiths in the town, " that use to make knives
and all manner of cuttynge tooles and many lorimars that
make bytes, and a greate many naylors, so that a great parte
of the towne is mayntayned by smithes." Camden found
" Bremicham " swarming with inhabitants, and echoing with
the noise of anvils from the vast number of smiths. The lower
part was marshy, but higher it was covered with an abundance
of handsome buildings. It continued a centre for the production of iron goods until the time of Charles II., when the iron
was worked up in open shops facing the street ; a few of which
existed in " Digbeth " till 1783. Hutton, journeying from
Walsall into Birmingham in 1741, could not conceive how
even so populous a country could support so many people of
the same occupation, and " was surprised at the prodigious
number of blacksmiths' shops upon the road." Walsall, though
but a small market town, possessed a Merchants' Guild in 1390,
and on Leland's visit held many smiths and bit-makers, and
he speaks of its iron ore and pits of sea coal. Defoe found spurs,
bridle-bits, stirrups, chapes, and shoe-buckles being manufactured, and until the latter fell into disuse this trade flourished.
Birmingham became exceedingly populous in the latter half of
the eighteenth century, exceeding Nottingham or Manchester,
and exported to all parts of the world. Prosperity greatly
increased as gun-making became added under William III.,
guns having previously come from Holland. The making of
steel by the Kestle family, in the seventeenth century, also gave
it an impetus, Steel House Lane marking the site until 1797.

In 1757 Birmingham supported the importation of iron from America, on the ground that England did not produce half that required, and that other imports would then cease; but it petitioned against Americans erecting slitting and rolling mills, for fear of injury to our own manufactures. Parliament decreed this prohibition, including the making of steel. Birmingham continued to excel in iron-working, the manufacture of brass not commencing much before 1740. In Wolverhampton lock-making is an old-established business, and as far back as the middle of the eighteenth century the town locksmiths were reputed to be the most ingenious in England.

The antiquity and importance of pin-making is somewhat surprising. Fuller speaks of it in connection with Gloucester, where it later became a monopoly employing 400 hands. Pins had before this been imported to the extent of £60,000 per annum. Defoe says that the first mill for slitting bars for wire-making was probably the one at Dartford. Beckman's account is that all wire was made by hand in England till 1565, when drawing it was introduced from abroad. Tintern, however, secured patents under Elizabeth and erected mills and held the trade for years. Wire-making was also established at Esher; and flatting mills at Sheen, by Dutchmen, in 1663. Leland mentions that many pinners were at Sherburn, but Aberford, in Yorkshire, later became famous for pin-making, its productions being in particular request by the ladies of the days of William and Mary.

In the South, Salisbury, Godalming and Tonbridge were each for a time noted for cutlery and steel-work. Woodstock employed some thirty hands on fine steel watchguards and other polished work, the workmen, according to Defoe, earning from 15s. 6d. to 42s. per week, when agriculture brought but from 6s. to 9s. Camden writes of a mill at Marlow, erected for the making of thimbles, hitherto imported, but " now become the manufacture of England to our advantage and reputation "; also of tin-plate works at Bisham, " which before had come from foreign parts."

London, as the Metropolis, was always a city of trade rather

M 2

than manufacture, though craftsmen of every kind formed an important element of its population. At all times armourers and smiths were busy within its walls, for weapons were ceaselessly being perfected and adapted to new means of attack. Iron was in demand for great engines of war, such as the movable towers with battering rams ; the catapulta, balista, and onager were shod and bolted with iron, while the defences of a first-class castle, such as Devizes, set up by Bishop Roger, Chancellor and Treasurer in the time of Henry I., might have six or seven portcullises, and these, perhaps, *ex solido ferro*, but in all cases shod with iron. The importance of the portcullis was great, for it is found as part of the arms of upwards of thirty towns, and adopted even by royalty as a cognisance. The Lancaster Portcullis, in gold, was used in his arms by its first Earl, Edmund. Considerable use of iron was made in the no less indispensable drawbridges. There were also iron gates, beacons, grates, grids, and bars, bolts and hinges, chains, and weapons and armour without end, horse-shoes, bridles, spurs, knives, and other trappings and harness, parts of implements of agriculture, and tools and nails.

Farriers must have been working in London at least from the Norman Conquest, and a sort of lustre was shed on them by Robert de Ferrariis, first Earl of Derby, whose shield was bordered with horse-shoes. They were first incorporated with cutlers and smiths into a single " mystery " by Henry III., who regulated their prices, and imposed a duty of a farthing per cwt. on iron, whether imported or exported from London. Ironmonger Lane became so called at that time from the members of the craft who worked there. On complaints of short measure from the wheelwrights, the King caused specimens of iron for cart-wheels brought from the Weald to be sealed by the Corporation, while the ironmongers " of the bridge " and " market " were to warn carriers that their wares, unless accurate to pattern, would be forfeited. Edward II. increased the dues to a penny per cwt., or 20d. per ton, when the King's beam was erected at the water-side. The Steelyard was instituted for merchants of Almaine who sold steel. The City customs, in addition to the King's, varied from 2d. for iron bars to 4d. for nails and manufactured steel articles imported from Germany. He sent his smith, Davide de Hope, to Paris to learn to make swords. Edward III. imported raw and manufactured iron, steel and wire, and in 1354 forbade its export, and again more strictly in 1377, especially wrought iron shot and guns. Flemish ironworkers were already well established in London, and his own armourer, John de Cologne, was a German. In this reign blacksmiths first became makers of large iron clocks, then being introduced by the building of a tower for the great clock at Westminster, for which the bells alone weighed 20,000 lb.

The interminable wars with France, and civil wars following the forcible dethronement and death of Richard II., kept the smiths busy, for they had to accompany the forces in the field, if not actually taking part in the *mêlées*, thus leaving little to chronicle as to their work at home. Under Henry IV. the smiths who forged the blades, the makers of knife-handles, and those who made sheaths for swords and daggers drew together and were incorporated as the Cutlers' Company by Henry V., and confirmed in the fourth year of Henry VI., who also incorporated the armourers, and is said to have become a brother of the Company. The Ironmongers' Company was first granted arms in 1453. Stowe says that the knives made in England in his time were coarse and uncomely, and that others were therefore brought in ship-loads from Flanders. One of the satisfactory sources of revenue of Henry VIII. was squeezing the German Steelyard, in return for concessions : but he also showed a preference for foreign armourers, founders and smiths, favouring impartially French, Flemish, German, Italian or Dutch. Some were installed to make armour in Greenwich, and others in Southwark, which later became a seat of ironfounding. Founders were here to cast large bronze cannon in the Tower, but either they failed to keep pace with Henry's immense requirements, or the great expense of the metal hampered them. An astute ironfounder of the Weald therefore proposed to cast guns for him in iron, and some of his foreign founders were promptly sent down to help. Iron guns were previously laboriously wrought of bars banded together at great expense. They succeeded beyond expectation, and laid the foundations of a great monopoly industry which lasted for over a century. Nails and locks at this time came to us from Holland, and armour in great quantities from Germany and Italy. Edward VI. removed the Greenwich armour factory to London, and Matthew Derick petitioned for permission to set up a store in London to teach the making of armour. He also encouraged importation, and the Swedes were favoured by paying no more dues for imported steel than those charged on our native production. The nice question of foreign *versus* English production was left for Elizabeth. She seemed to favour her father's policy, for the German Steelyard maintained its practical monopoly almost to the end of her reign. Pennant describes it as the great repository of imported iron which furnished the Metropolis with that indispensable material. " The quantity of bars that fill the yards and warehouses of this quarter strike with astonishment the most indifferent." Next the water-side were fixed two masts with the imperial eagles, and the Steelyard was practically a foreign stronghold, under its own rules, in the heart of London. Elizabeth was the first to disapprove and check the wholesale destruction of our forests as fuel for iron and glass furnaces. It must be remembered that forests and timber then stood for all that steel, iron and coal are now to us. Timber especially meant ships and houses and fuel.

Queen Elizabeth seems to have possessed as true, capable, and wide an outlook as any monarch of them all, and probably it appeared to her better, at all costs, to check the destruction of so valuable a national asset, on which our national life seemed then to depend. She thought imperially, for Drake had just made his wonderful voyage. It had, indeed, been urged in 1556 under Queen Mary that English mills should be entirely closed, as producing inferior qualities of steel for armour. Thus Harrison wrote that iron could be brought from abroad better and cheaper than it could be made at home. Also, " that our steel is not so good for edge tools as that of Cologne, yet one is often sold for the other," and that iron would be cheaper if all was imported. At the same time, Spanish iron had been brought down to only five marks per ton, while English iron was nine marks. By 1596 the Germans felt it safe to increase their prices to £14 for iron and £15 for steel, when Elizabeth at last, in 1598, decided to strike her heaviest blow, the final extinction of the German Steelyard, which had maintained itself in our midst for centuries. Not many years before, in 1585, all Spaniards, and some Italians and Dutchmen, had their goods seized for Her Majesty's use. Meantime, Flemings had been flocking into England to escape persecution, and a failure in the banks of the Witham, near Boston, brought over a Flemish engineer with novel machinery, mechanics and iron. Another was permitted in 1582 to erect engines in an arch of London Bridge to pump water from the Thames to supply it to London. These led, early in the next reign, to the introduction of portable pumping engines to extinguish fires. Elizabeth now did all she could to encourage English workers. The smiths and blacksmiths had been finally incorporated into a company in 1578, and Londoners were encouraged by the grant of a monopoly of the export of all goods made in Sheffield, then becoming the principal mart for the finer qualities of cutlery.

A Londoner of Fleet Bridge, Richard Matthews, was, it seems, the first Englishman to obtain great skill in making knives and their hafts, being granted a prohibition against all strangers and others from bringing any knives into England from beyond the seas. The art of making fine Spanish needles had been brought here by a negro in Mary's time, but died with him. Under Elizabeth a German, Elias Crowse, taught this as well as pin-making, and the industry became firmly established. Magnificent armour was produced in Greenwich under Elizabeth, but whether by a German master is not quite clear. It seems significant that in the muster of 1585 at Greenwich no armourers or blacksmiths appear, though twenty-seven cutlers and 147 ironmongers took part. The Minories, which had been inhabited by gun-smiths, a fact alluded to in old plays, had been turned into large store-houses for armour and habiliments of war.

The succession of James made little difference. Very fine suits of armour exist, made for his sons Henry and Charles when they were striplings. A rare book gives the cost of an outfit of russeted armour for a cuirassier at 90s., and for a foot-man 22s. in 1610. The latter comprised a back and breast plate and head-piece. The art of tinning iron, invented in Bohemia, was carried as far west as Saxony in 1620. Smithcraft made no perceptible progress except in the relatively trivial window casements and door hinges and fastenings of new and curious forms, quaint and in great variety.

Charles I. was heavily handicapped by the out-of-date precepts of his pedantic father, and had no real ideas as to the awakened and advanced outlook of his subjects. We see this in Lord Holle's " England is a kingdom of territory not of trade," a rebuke to the Dutch Ambassador, so applauded by Strype that he conceived " it should be set in golden characters and preserved to all posterity." Regal patronage brought flocks of foreign artists to his court, who were preferred, except perhaps in architecture, to natives. Among his treasures was a gold ring, with lion and unicorn and royal arms marvellously and exquisitely carved in steel in high relief on the bezel. In the Tower were 100,000 suits of armour, some gilded and engraved to perfection, and 2,000 cannon lay on the quays with supply of ball. The gathering troubles were foreshadowed when the Duke of Rutland learned that the armourers of Greenwich were so busily employed that he could get no armour there. In the year of Charles's death coals in London were 30s. per chaldron, with fears of a rise to £3 before the winter was over.

The policy of encouraging native crafts was revived by Charles II. Tin-plate and wire works were incorporated, Germans being brought over to establish the former, while a Dutchman established at Sheen the first flatting mill seen in England. Iron-plate mills were set up at Wimbledon, and foundries at Southwark, and in 1695 a manufactory of thimbles by a Dutch-man at Islington. The rate paid at this time at the King's beam for iron entering the City was 20d. per ton, while the City toll for bar iron was 2d., and manufactured wares, including ordnance, 4d. Steel came in bundles or small barrels, which paid 2d. Our exports were then iron, wrought and unwrought, shot-guns, knives and scissors, the latter going to India. Our imports were iron, steel, wire, sword and knife blades. The ironmongers had overflowed from their " Lane " to Old Jewry and Thames Street, while Pope's Head Alley became chiefly for cutlers. The consumption of coal in the City was now becoming regarded as a nuisance. In 1684, during a heavy frost, Evelyn complains that one could hardly see across the streets and was choked by the smoke. Just before the Great Fire, it was enacted that provision of coal was to be held by the City Companies. Just after, the price went up to £3 3s., and within six months to £5 10s., according to Pepys. Coal, " sea," " pit," and Scotch, was brought into the Thames from Newcastle, when the woodmongers went far down to intercept it, though prohibited under forfeit of 5s. per ton. Four hundred and twenty carts were at this time licensed to carry it, and paid £820 per annum for the privilege, beside the duty of 2d. The duty put on it later brought in, between 1716 and 1724, sufficient to build fifty new churches, to grant £4,000 for repairs to Westminster Abbey, and £6,000 for Greenwich Hospital. Among the many bubbles of 1720 was the making of tin-plate. For a century or two before the eighteenth century the import of iron from Sweden had been considerable, until at last it seemed to menace our prosperity, the imports from thence totalling more than those of all the rest of Europe put together. Meanwhile our American colonies were restricted by severe duties from exchanging their iron for other commodities with the Mother Country. A Parliamentary enquiry in 1750 repealed the duties, but forbade the establishment in America of mills for slitting or rolling. Imports from Sweden and Russia continued, and bar iron from America was imported free, but not to be carried coastwise or more than ten miles by land, there then being 109 forges in England and Wales, producing 18,000 tons, besides Scotland. It was said at this time that our mines were inexhaustible, and but for American importation duty free, our output would greatly increase. Works here consumed 198,000 cords of wood, produced on lands which it was said would otherwise be barren. Swedish iron was superior for steel-making, and American iron unfit for edged tools or ship requirements. Sheffield urged that their work would be undersold. Birmingham contended that England could not supply half our requirements, and advocated American imports, but without permission being granted to them either to roll or split.

The French War having crushed that country's navigation, our traders supplied foreign markets on their own terms, where before they were undersold and met with dangerous competition. Our export trade, augmenting to a surprising pitch, alone enabled us to maintain our fleet at an enormous expense. But this expansion of trade, and the avarice and rivalry of contractors and middlemen in cutting prices, led to bad work. Thus many of our steel and iron productions fell into discredit abroad. Razors, knives, scissors, hatchets, swords and other edged tools made for exportation, were generally badly tempered, half-finished, flawed, or brittle.

THE WEALD.

THE WEALD, old English for forest, extends over parts of Kent, Surrey and Sussex, its soil consisting chiefly of mottled clay varied by sandstone, in which certain layers are ferruginous or rich in iron oxide. Geologically it represents an ancient river delta laid bare by subsequent denudation of the overlying chalk and newer strata. The nature of the ores, previously unknown, must have been discovered by the Romans, since Roman remains have been found at Maresfield, Chiddingly, Sedlescombe, Westfield, Horsham, and near Hastings. These relapsed to the pristine state of impassable swamp and forest on their departure, and so remained for upwards of a thousand years. A cast iron grave-slab is the only indication of a revival of the industry in the fourteenth century, but wrought iron may well have been produced in the Weald at a considerably earlier period. In those times iron was produced from the ore in a furnace, with bellows of leather and wood ; while to convert the " bloom " into merchantable iron needed further process in a forge, the hammers employed to convert it into bars being " tilt " or lever worked by water power. The furnace and the forge were separate buildings and establishments, and are invariably kept distinct in documents, even though worked by one owner : who might possess two furnaces and one or more forges. Diggers, carters, ostlers, wood-cutters, charcoal-burners, stokers, puddlers, moulders, smiths, mates, in all some sixty-five men, were required for a furnace and forge, of which there were at least 150 or more in the Weald in the days of its prosperity. The bar iron fit for blacksmith or wheelwright was carted chiefly to London, but a large wholesale trade in finished nails, horseshoes, arrow-points and such-like also existed. Cannons and mortars of bars dovetailed and strongly banded together were constructed on the spot. By some fortunate chance two of these have escaped destruction : a large breach-loading mortar, capable of throwing stone balls of 160 lb. weight, was found in the moat of Bodiam Castle ; the other, a long gun, preserved on Eridge Green, is similarly constructed. Both were found to have inner linings or inner chambers skilfully cast in iron, showing a degree of technical skill hardly to have been looked for in the early years of the fifteenth century, and unknown abroad.

Of cast iron shot there is no actual evidence in England prior to the reign of Henry VIII. ; but it is related that the Master of the Ordnance to Henry VII. was shown in France a shot of " yeron 28 ynches aboute " for a siege piece. The difficulty and cost of forging solid shot and, still more, bombs, made casting essential at the earliest moment, and Henry VIII. induced his ally, the Emperor Charles V., to send him the best founder of cannon shot in Spain. On the eve of his French and Scottish wars, in 1543, Henry interviewed Ralph Hogge, an ironfounder of Buxted, who covenanted to cast cannon of iron at the rate of £10 per ton, as against £70 paid for bronze. Henry thereon sent his most famous bronze gun founders, Peter Baude, a Frenchman ; Arcanus de Cesena, an Italian ; Van Cullen, probably Dutch ; and the English Owens and Johnsons, to render assistance. With their help Ralph Hogge erected a cannon foundry. The Owens also enabled him to cast shot, both solid and hollow. In 1549, the King possessed the furnace of Worth, an inventory including six tons five cwt. of cast iron shot for culverins, and thirteen tons for guns of other calibres ; twenty-eight hands were employed, with two gun founders and a master of the ironworkers. Henry, from personal interest, also controlled the cutting of timber, the supply being inadequate to the increased demand. In 1552, on the impeachment of the Lord High Admiral Seymour for treason, the Horsham mills were confiscated by Edward VI. On Elizabeth's accession the indiscriminate felling of timber was prohibited, and later, in 1574, every furnace or forge owner was required to give bonds. Finally Elizabeth and her far-seeing statesmen forbade the building of any more iron works in Surrey, Kent, or Sussex. She herself owned and sub-let a furnace and forge at Maresfield, near Ashdown Forest, and also those of Worth and Horsham. Among notable owners have been the Dukes of Norfolk and Northumberland, and the Lords Abergavenny, Ashburnham, Buckhurst, Dacre, Derby, Montague, Monteagle, Sackville, Seymour, and Surrey ; together with knights and gentry including the Pelhams, Sidneys, and other of Elizabeth's courtiers. In all, 179 names have been recovered, and forty other iron-masters are known by initials, some of whom have been already mentioned.

The seventeenth century opened prosperously for the Weald, and under James I. half the iron trade of England still remained with Sussex. The possibility of smelting iron with coal was, however, already in the air, and in 1611 a patent for it was taken out by Sturtevant. More foundries and forges were now in royal hands, but this was perhaps hardly an unmixed benefit to general trade. An instance may suffice to show the existing conditions. Among the Crown properties was St. Leonard's Forest, in which Elizabeth held two forges and a furnace in 1574. These came into King James's hands and were leased in 1608. Fuel was becoming scarce, yet Charles I. granted this furnace in 1630 the right to dig ore, and to 250 cart-loads of charcoal and thirty loads of wood for fuel. This may have been a fair amount, but restrictions were certainly necessary, though perhaps they became irksome and even unfair. Cromwell had sufficient power in 1643 to order the destruction of all the royal furnaces, and sent Waller into Sussex to see this carried out. In the following year the forges also, not only those belonging to or leased by the Crown, but all those in the hands of malignants, were to be destroyed. The orders were ruthlessly carried out in Waller's accustomed manner, and the important works in St. Leonard's Forest, and those at Horsham, Burwash, Buxted, Ewhurst, Rudgwick, and the rest ceased to exist. After such drastic and high-handed proceedings, there remained but twenty-seven furnaces in the Weald, their combined output but 1,400 tons. So much royal ownership in competition may have formerly led to abuses, for Cromwell's acts were rarely senseless, and his revenue suffered considerable loss. For the next twenty years the number of furnaces and forges continued to diminish, but Anthony Morley, of Horsted Keynes, rescued as much as he could of the Sussex industry by carrying it into Glamorganshire. The number of furnaces then further diminished to ten. But these comprised some of the largest and most important, all being subsidised to produce cannon and shot for Government under Charles II., who patronised and protected them. They could not compete in price with the coal-smelted iron of the Midlands, where iron-masters were enabled to adopt all the newest inventions, such as iron bellows, grooved rolls for making bars, etc. The quality of iron smelted by charcoal, however, remained undoubtedly superior.

The eighteenth century practically saw the end of the iron industry of the Weald. Dud Dudley had patented the use of coal for smelting iron, and in 1735 a further improvement was effected by Abraham Darby smelting iron for casting with coke instead of coal in high furnaces. In 1728, Payne and Hanbury invented rolling mills for sheet iron. Cylindrical iron bellows were introduced in place of wood and leather by John Smeaton in 1760, and grooved rolls for bars by Cocts in 1783; while in 1784 cast iron was converted into wrought. The Weald furnaces were, notwithstanding, supported for some time longer

by Government subsidies, and all continued to be employed
in making cannon till 1740. Their charcoal iron was equal to
Swedish and entirely free from sulphur, and hence preferred.
The last furnace to receive the Government subsidy was Fern-
hurst, in West Sussex, not closed till about 1770. Robertsbridge,
at one time owned by the Sidneys, did not close till 1768.
Heathfield shipped the best cannon from Newhaven. Worth
furnace continued to produce and send its cannon to London
till 1780. By 1778 only two others remained, Maresfield and the
Ashburnham forge, and produced no more than 300 tons
between them. In 1796 only the latter remained, with an output
of 173 tons, lingering till 1828. During the later days, no doubt,
such trivialities as rush sticks, tinder-boxes, toasting forks,
and other domestic necessities dear to collectors were produced
for local sale.

GENERAL INDEX.

TOPOGRAPHICAL INDEX.

O

SUPPLEMENTARY BIBLIOGRAPHY.

A) Wrought Iron

AMATT L. K. — *Locks & lockmaking – an annotated bibliography*, London, 1973.

AYRTON, M. & SILCOCK, A. — *Wrought iron and its decorative use*, London, 1929.

FERRARI, G. — *Il ferro nell'arte italiana*, Milan, 1924.

GEERLINGS, G. K. — *Wrought iron in architecture*, London, 1929.

HARRIS, J. — *English decorative ironwork 1610–1836*, London, 1960.

HOLLISTER-SHORT, G. J. — *Discovering wrought iron*, Shire Publications, 1970.

LISTER, R. — *Decorative wrought ironwork*, London, 1957.

SCHUBERT, H. R. — *History of the British Iron & Steel Industry, c.450 BC – 1775 AD*, London, 1957.

SEYMOUR-LINDSAY, J. — *An anatomy of English wrought iron*, London, 1964. *Iron and brass implements of the English house*, London, 1964.

SONN, A. H. — *Early American wrought iron*, New York, 1928.

STARKIE GARDNER, J. — *English ironwork of the 17th & 18th centuries*, London, 1911.

B) Cast Iron

CHATWIN, A. — *Cheltenham's Ornamental Ironwork*, Cheltenham, 1975.

GLOAG, J. — *Cast iron in architecture*, London, 1948.

KAUFFMAN, H. J. — *Early American ironwork, cast & wrought*, Tokyo, 1966.

LISTER, R. — *Decorative cast ironwork in Great Britain*, London, 1960.

MAINWARING BAINES, J. — *Wealden firebacks*, Hastings, 1958.

STRAKER, E. — *Wealden iron*, London, 1931.

TRINDER, BARRIE — *The Darbys of Coalbrookdale*, Phillimore Press, 1974.

C) *Exhibition and Museum Catalogues*

France, Rouen, Le Musée Le Secq des Tournelles: *Ferronnerie ancienne* by H. R. D'Allemagne, Paris, 1924.

London, Burlington Fine Arts Club: *Exhibition of Steel & Ironwork 1900.*

London, Ironmongers' Hall: *Exhibition of Antiquities & Works of Art 1861.*

USA, Houston, University of St Thomas: *Exhibition: Made of Iron,* Sept – Dec 1966.

PLATES.

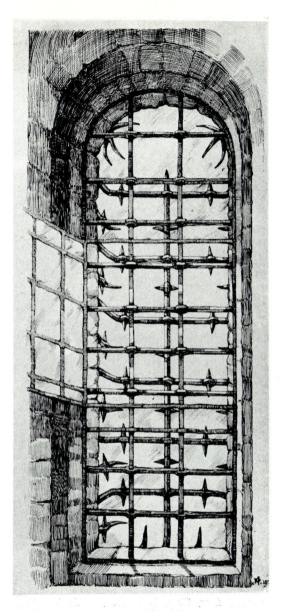

Fig. 4.—Treasury window-grille, Canterbury
Cathedral. (p. 16).

FIG. 5.—South door, Skipwith Church, Yorkshire. 12th century. (p. 18).

Fɪɢ. 6.—St. Swithin grille, Winchester Cathedral. End of 11th century.
(p. 19).

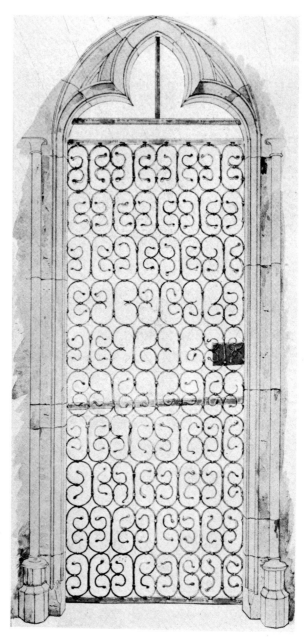

Fig. 7.—Grille of St. Anselm's Chapel, Canterbury Cathedral.
About 1100. (p. 19).

FIG. 8.—Part of choir gate, Canterbury Cathedral. Beginning of 14th century. From a reproduction in the Victoria and Albert Museum, No. M 1912-2. (p. 21).

FIG. 9.—Door of Merton College, Oxford. Middle of 13th century.
(p. 24).

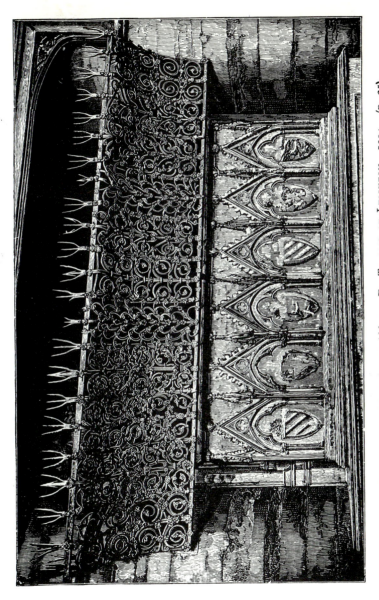

FIG. 10.—Eleanor grille, Westminster Abbey. By THOMAS OF LEIGHTON. 1294. (p. 25).

FIG. 11.—Iron-mounted cupboard. From Whalley Abbey, Lancashire. 14th century. In the Victoria and Albert Museum, No. M 170–1917. (p. 26).

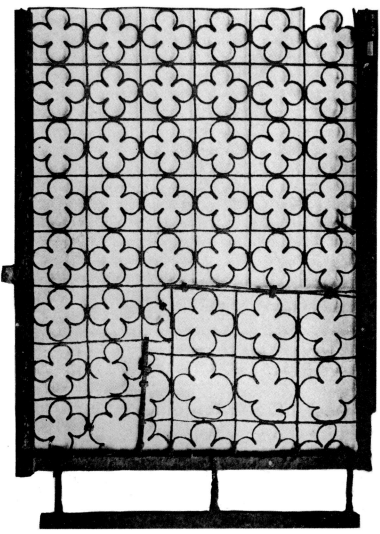

Fig. 12.—Part of a pair of gates from Chichester Cathedral.
Second half of 15th century. (p. 29).
In the Victoria and Albert Museum, No. 592-1896.

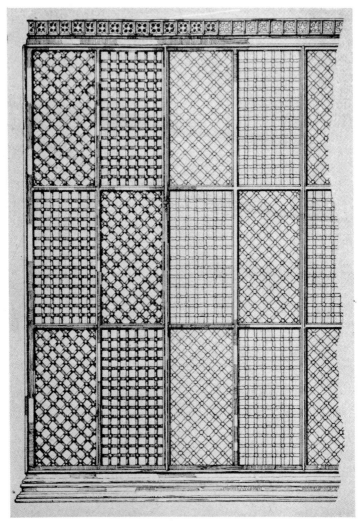

FIG. 13.—Grille, St. Albans Abbey. Middle of 15th century. (p. 30).

FIG. 14.—Grille, Henry V. Chantry, Westminster Abbey. 1428.
(p. 31).

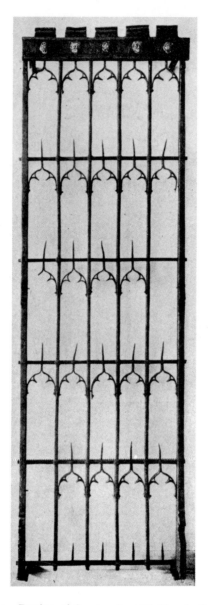

Fig. 15.—Portion of Screen, Arundel Church, Sussex.
Early 15th century.
From a reproduction in the Victoria and Albert
Museum, No. M 1911-8. (p. 31).

Fig. 16.—Tomb-railing (Herse). From Snarford Church, Lincolnshire. 15th century. In the Victoria and Albert Museum, No. 47–1867. (p. 32).

FIG. 17.—Upper part of a gate pier, St. George's Chapel, Windsor. By JOHN TRESILIAN. End of 15th century.

(p. 34).

FIG. 20.—Gates of Bishop West's Chapel, Ely Cathedral.
Made in 1515. (p. 39).

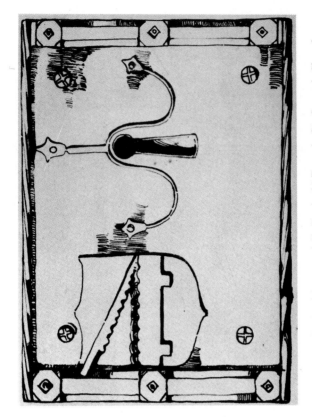

Fig. 19.—Lock with arms of Sir Reginald Bray, in St. George's Chapel, Windsor.   About 1500.   (p. 37).

Fig. 18.—Lock of Bishop Edmund Audley, in Hereford Cathedral. End of 15th century.   (p. 37).

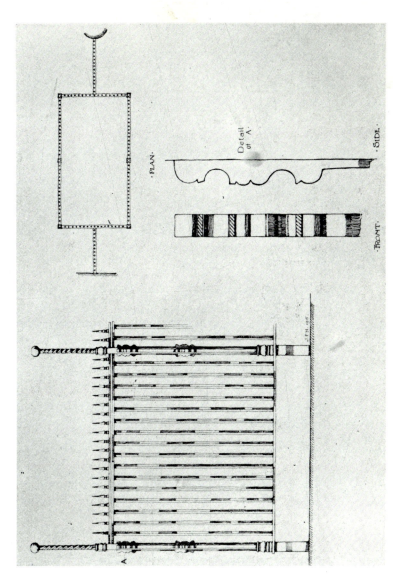

Fig. 21.—Railing to the Pickering Monument in St. Helen's, Bishopsgate. 1575. (p. 46).

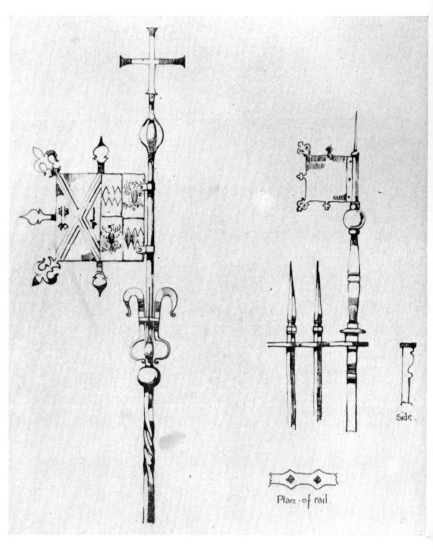

FIG. 22.—Standard from railing of
Bishop Montague's tomb in Bath
Abbey. 1618. (p. 49).

Railing of Hoby Monument in
Bisham Church, Berkshire. Second
half of 16th century. (p. 49).

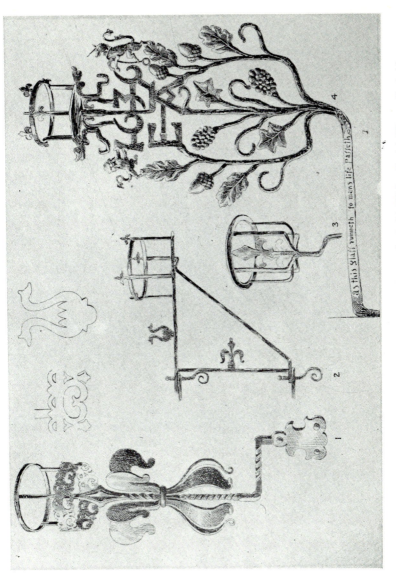

FIG. 23.—Hour-glass stands at (1) Linlithgow, (2) Stratford, (3) Kirkwall, (4) Hurst, Berkshire. 17th century. (p. 56).

FIG. 24.—Suspension-rod for a chandelier. Second half of 17th century. In the Victoria and Albert Museum, No. 170–1865. (p. 73).

Fig. 25 —Cresting from a screen at Hampton Court Palace. By Jean Tijou. End of 17th century. (p. 80).

Fig. 26.—Part of screen at Hampton Court Palace. By Jean Tijou.
End of 17th century. (p. 81).

FIG. 27.—Panel of sanctuary screen, St. Paul's Cathedral.
By JEAN TIJOU. End of 17th century. (p. 88).

FIG. 28.—Garden screen at New College, Oxford. By THOMAS ROBINSON. Completed about 1711. (p.95).
Reproduced from "English Ironwork of the XVIIth and XVIIIth Centuries,"
by J. Starkie Gardner; published by B. T. Batsford, Ltd.

FIG. 29.—Screen from St. John's Church, Frome. Perhaps by PARIS of Warwick. Beginning of 18th century. In the Victoria and Albert Museum, No. 1092–1875. (p. 100).

FIG. 30.—Gates at St. Mary's Church, Oxford. Attributed to PARIS of WARWICK. Beginning of 18th century. (p. 101).
Reproduced from "English Ironwork of the XVIIth and XVIIIth Centuries,"
by J. Starkie Gardner ; published by B. T. Batsford, Ltd.

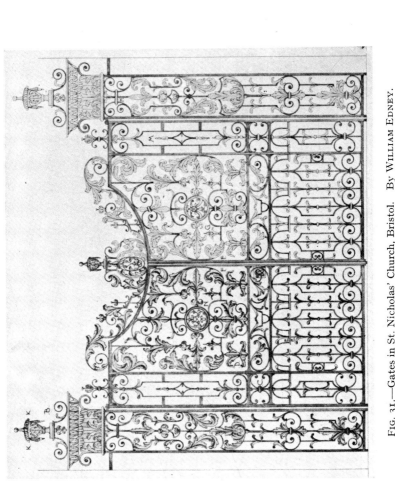

Fig. 31.—Gates in St. Nicholas' Church, Bristol. By William Edney. Early 18th century. (p. 104).

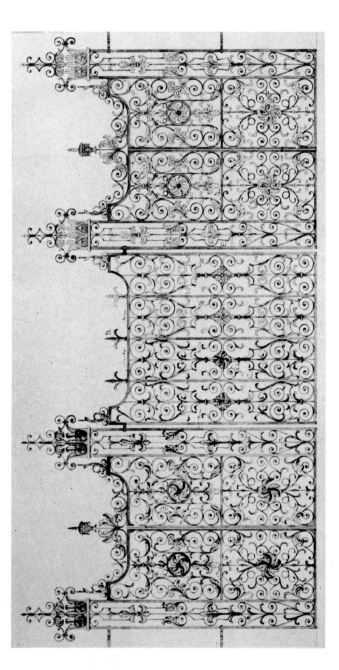

Fig. 32.—Screen in Temple Church, Bristol. Made by William Edney. 1726. (p. 104).

Fig. 33.—Railing and gates at Chirk Castle, Denbighshire. By Robert Davies. 1715–21. (p. 105).
Reproduced from "English Ironwork of the XVIIth and XVIIIth Centuries,"
by J. Starkie Gardner; published by B. T. Batsford, Ltd.

FIG. 34.—Porch to a garden house at Melbourne, near Derby. By
ROBERT BAKEWELL. 1707–11. (p. 108).
*Reproduced from " English Ironwork of the XVIIth and XVIIIth
Centuries," by J. Starkie Gardner ; published by B. T. Batsford, Ltd.*

FIG. 35.—Screen in All Saints' Church, Derby. By ROBERT BAKEWELL. 1723–5. (p. 109).

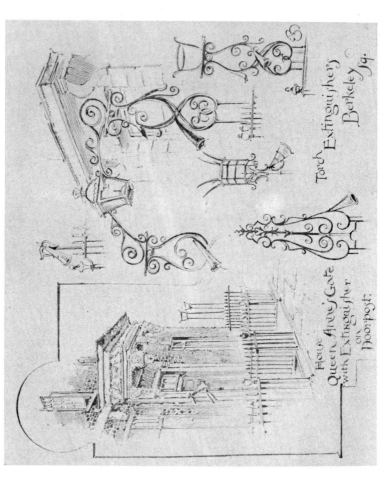

Torch Extinguishers Berkeley Sq.

House Queen Anne's Gate with Extinguisher on Doorpost.

Fig. 36.—Lamp brackets. 18th century. (p. 121).

Figs. 37, 38.—Brackets. 18th century. In the Victoria and Albert Museum, Nos. M 342–1910, 258A–1896. (p. 122).

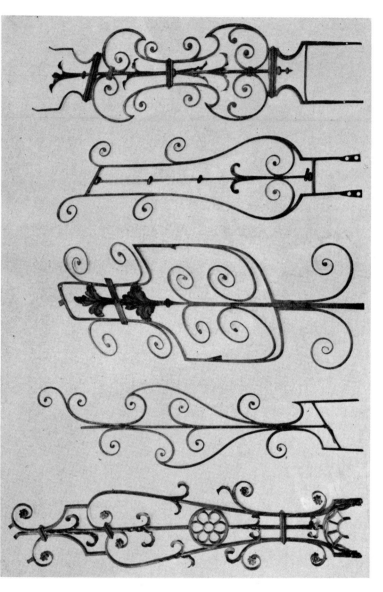

Fig. 39.—Balusters. 18th century. In the Victoria and Albert Museum. (p. 124).

Fig. 40.—Gates in the Green Park, Piccadilly, formerly at Devonshire House. Early 18th century. (p. 125).
Reproduced from " English Ironwork of the XVIIth and XVIIIth Centuries,"
by J. Starkie Gardner; published by B. T. Batsford, Ltd.

FIG. 41.—Gates at Trinity College, Cambridge. Early 18th century. (p. 127).
Reproduced from " English Ironwork of the XVIIth and XVIIIth Centuries," by J. Starkie Gardner ; published by B. T. Batsford, Ltd

FIG. 42.—Gate. About 1750. In the Victoria and Albert Museum, No. M 278–1920. (p. 132).

Fig. 43.—Suspension-rod for a chandelier. First half of
18th century. In the Victoria and Albert Museum,
No. 646–1888. (p. 136).

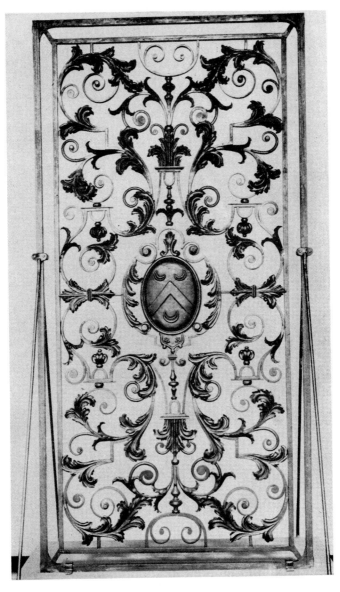

FIG. 44.—Panel of a gate from Bridewell Hospital. 1714.
From a reproduction in the Victoria and Albert Museum,
No. 1888-424. (p. 145).

Fig. 45.—Balustrade, formerly at 35, Lincoln's Inn Fields. Early 18th century. In the Victoria and Albert Museum, No. M 56-1921. (p. 146).
Reproduced from " The English Staircase," by W. H. Godfrey ; published by B. T. Batsford, Ltd.

Fig. 46.—Balustrade (upper part), formerly at 35, Lincoln's Inn Fields. Early 18th century. In the Victoria and Albert Museum, No. M 56-1921. (p. 146).

Reproduced from "The English Staircase," by W. H. Godfrey ; published by B. T. Batsford, Ltd.

Fig. 47.—Gate at Fenton House, Hampstead. About 1706. (p. 146).

Fig. 48.—Gate at Wovington Manor, Sussex.   Early 18th century.
(p. 146).

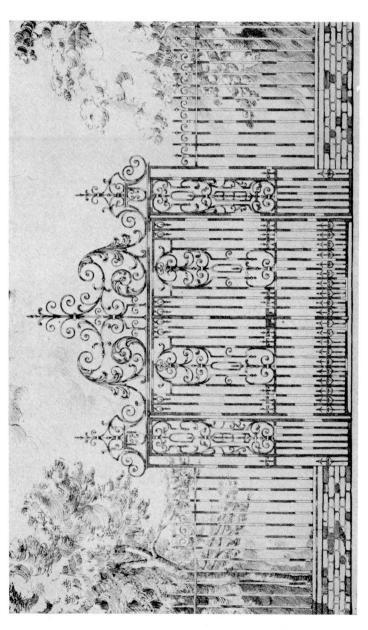

Fig. 49.—Gate and railings at Hampstead Parish Church.　First half of 18th century.　(p. 147).

FIG. 50.—Fanlight from Drapers' Hall, London. Designed by ROBERT ADAM. Late 18th century. (p. 156).

FIG. 51.—Fire-back. 15th century. In the Victoria and Albert Museum, No. 896–1901. (p. 159).

FIG. 52.—Fire-back. First half of 16th century. In the Victoria and Albert Museum, No. M 120–1914. (p. 160).